Development of the Iranian Oil Industry

Fereidun Fesharaki

The Praeger Special Studies program — utilizing the most modern and efficient book production techniques and a selective worldwide distribution network—makes available to the academic, government, and business communities significant, timely research in U.S. and international economic, social, and political development.

Development of the Iranian Oil Industry

International and Domestic Aspects

Praeger Publishers   New York   Washington   London

PRAEGER SPECIAL STUDIES IN INTERNATIONAL ECONOMICS AND DEVELOPMENT

Library of Congress Cataloging in Publication Data

Fesharaki, Fereidun.
    Development of the Iranian oil industry.

    (Praeger special studies in international economics
and development)
    Bibliography: p.
    1. Petroleum industry and trade--Iran--History.
2. Iran--Economic conditions. I. Title.
HD9502.I72F47   338.2'7'2820955   75-19782
ISBN 0-275-55600-X

PRAEGER PUBLISHERS
111 Fourth Avenue, New York, N.Y. 10003, U.S.A.

Published in the United States of America in 1976
by Praeger Publishers, Inc.

Printed in the United States of America

To My Wife and Parents

When a modern analyst thinks about Iran as an integral entity and considers her history, cultural and intellectual heritage, he will necessarily realize that what characterizes Iran is permanence amidst change. At times, as now, this change is dynamic and constructively directed toward the realization of permanence itself. What is characteristic of Iran, as expressed in her art, culture, and thought is the interrelationship between her manifest change and dynamism on the one hand and her essential permanence and stability on the other.

The birth of modern Iran coincided with the coronation of Reza Shah Pahlavi "the Great" in 1925. Under his rule great strides were made toward unification of the country, laying the infrastructural basis for future development of the economy and social reforms. More importantly, he built the foundation for a national policy, independent of the foreign powers who had dominated Iran for the past century.

His son, the present Shah--H.I.M. Mohammad Reza Pahlavi Aryamehr--came to the throne in 1941. Under his progressive leadership the Iranian oil industry was transformed from a foreign-controlled industry to a modern and sophisticated national entity. On the international scene, his leadership of the Organization of Petroleum Exporting Countries (OPEC) resulted in an increase in the Iranian oil revenues from $34 million in 1954 to $18,000 million in 1974--an increase of over 520 times--and an OPEC surplus of $50-60 billion in the same year. His decisions on the oil scene have, in the past, had an important impact on the economies of the industrial countries. On the home front, the National Iranian Oil Company (NIOC) took over the distribution activities of the foreign operator in 1951 and expanded the domestic oil industry into a gigantic entity. While the oil revenues fueled the economy, the cheap and easily accessible oil products encouraged the expansion of the oil-based industries in the country. Except for an economic recession in the early 1960s, the economy has generally enjoyed consistent high rates of growth; 7 percent in 1959-70, over 14 percent in 1971-72, and between 30 to 50 percent in 1973-75.

Iran's role among the Middle East oil producers has been that of a pioneer. Iran is the oldest Middle East producer, the first to nationalize its oil industry, the originator of the national oil company concept in the region, the promoter of the new and innovative oil contracts, and not least, the most economically advanced oil producer in OPEC. Although this book focuses on Iran in particular,

vii

the discussions are kept within the context of the economics and politics of the oil industry in the region as a whole. Furthermore, it is hoped that the various issues discussed will throw light on some of the general problems of the international oil industry.

Part I is concerned with the economic and political implication of the early oil concessions in the period 1901-51. It also looks at the domestic oil market and the structure of supply and demand, if only to show the difficult tasks NIOC had to face in transforming the disintegrated market into a coherent one.

Part II discusses the process through which NIOC and the Iranian government succeeded in asserting Iran's control of its industry and the international implication of these activities. The Iranian nationalization of 1951, the Consortium agreement, and the award of nonconcessionary joint venture and service contracts are considered. A comparison of the conventional and nonconventional contracts is also carried out to show whether any type is preferable to the other on both political and economic grounds. In examining the role of Iran in OPEC in the past 15 years we consider the importance of the Iranian influence in the organization, the conflicts of interest between the members, and the chances of the survival of the organization in the future. The book also deals with the issue of oil revenues in Iran and examines their utilization through domestic expenditures and foreign investments. We will consider the so-called "petrodollar problem" from the point of view of the producing and consuming nations of the world and the particular role of Iran in the recycling of such funds.

Part III deals with the domestic activities of NIOC in Iran, a subject which has so far been ignored by the scholars. It discusses the objectives of the Company and the manner of their implementations. Thus exploration, production, distribution, and transport are studied. After examining the sources of supply we next look at the structure of the demand and the factors that have influenced this unprecedented growth in the past two decades. It will be seen that the patterns of supply and demand are not matched with resulting imbalances in the system. The causes of this imbalance are studied and possible remedies offered. Included also are price and cost structures and the profitability of the oil operations. In all cases we will try to show the influence of the government in the domestic and foreign operations of the Company, and will attempt to show the compromise between the financial aspirations of a huge oil company and the national policy of the government. As we shall see, because NIOC is a strange mixture of a giant business corporation and a government department, its activities cannot be judged on a purely commercial basis. With this in mind, a detailed study of NIOC's operations is carried out so as to identify not only the economic but also the political and social pressures that have determined its activities. Some remedies are offered to ease the impact of the noneconomic forces on the Company.

An important obstacle to research in developing countries is the lack of data. Iran of course is no exception. The data relating to the financial activities of the Anglo-Iranian Oil Company were made available by the British Petroleum Company, although its scope was limited to very brief annual reports. The systematic and regular collection and collation of statistics in Iran really started in the mid-1950s. In some cases cross-checking of the data showed that they were unreliable; in others the material was of a confidential nature and therefore not available. Thus it was unavoidable that the research focused on areas where information could be obtained. In this way the availability of the data has to a significant extent dictated the development of this study.

Published materials such as books and journals were used for the historical section and parts of Part II. For Part III there is little published material available. Most of the data on the domestic oil industry were collected during two field trips to Iran. In most cases original records were consulted; in others the records were not in a form that was immediately useful, and frequent instances of conflicting information had to be laboriously sorted out--often with a great deal of assistance from various experts. Sometimes it was found necessary to rely on the expert's opinion instead of statistical facts. Frequently the author was obliged to use his own discretion in selecting and interpreting the conflicting information.

The exchange rate (selling) was relatively stable from the 1960s until 1973 at $1 = 75-76 rials. In 1974 the exchange rate was $1 = 67.75 rials, but in March 1975 the rate changed to $1 = 66.77 rials, after the government decision to link the rial to the Special Drawing Rights instead of the U.S. dollar.

ACKNOWLEDGMENTS

This book is the outgrowth of a doctoral dissertation carried out at the University of Surrey in England. The present work--a fully revised and updated version of the original study--was completed during 1974-75 at the Center for Middle Eastern Studies of Harvard University.

I am indebted to more individuals than can be identified in this space. My greatest thanks are due to Dr. L. F. Haber of Surrey University and Professor E. T. Penrose of London University, who carefully read the original manuscript and made numerous suggestions for improvement of the work.

At the National Iranian Oil Company, I received valuable help from Dr. G. Naamati, Dr. M. Djafari, and Messrs. A. Aghdaii, R. Azimi, H. Azad-Peyma, N. Dorrifar, A. Farid, I. Delfanian, A. Shahlavi, and M. Nassiry. Mrs. F. Marvasti of the Bank Markazi Iran was particularly helpful.

I would also like to acknowledge the cooperation of the following, who either read parts of the manuscript, discussed various issues in the study, or in some other way helped me complete this work: Dr. Jamshid Amuzegar, Minister of Interior and Iran's Chief OPEC negotiator; Dr. S. Rasekh, Vice-Minister of Plan and Budget Organization; Dr. A. Kooros, Vice-Minister, Ministry of Economic Affairs and Finance; Professor M. Ganji, advisor to the Prime Minister; Dr. T. R. Stauffer and Professor R. B. Stobaugh of Harvard University; Professor C. Robinson of Surrey University; Dr. B. Dasgupta of the Institute of Development Studies at the University of Sussex; Dr. R. W. Ferrier of the British Petroleum Company, and Mr. K. Mochizuki of Nippon Steel Corporation and Harvard University.

None of the above mentioned are, of course, responsible for any errors that may be present in this work.

CONTENTS

LIST OF TABLES

xix

LIST OF MAPS AND FIGURES

LIST OF ABBREVIATIONS

AGIP   Subsidiary to ENI

AIOC   Anglo-Iranian Oil Company

APOC   Anglo-Persian Oil Company

APQ   Aggregate programmed quantity

BUSHCO   Bushehr Petroleum Company

CFP   Compagnie Francaise des Petroles

DAC   Development Assistance Committee

DD   Distribution Department

DOPCO   Dashtestan Offshore Petroleum Company

ENI   Enie Nazionale Idrocarburl (Italian State-owned oil company)

ERAP   Enterprise de Recherches et d'Activités Petroliers

FEC   First Exploration Company

FPC   Farsi Petroleum Company

GDP   Gross Domestic Product

GNP   Gross National Product

HFO   Heavy fuel oil

HOPECO   Hormuz Petroleum Company

IEA   International Energy Agency

IGAT   Iranian Gas Trunkline

ILO   International Labor Office

INEPCO   Iranian Nippon Petroleum Company

IPAC	Iran Pan American Oil Company
LAPCO	Lavan Petroleum Company
LPG	Liquefied petroleum gas
MSA	Most seriously affected (countries)
NIGC	National Iranian Gas Company
NIOC	National Iranian Oil Company
NITC	National Iranian Tanker Company
NPC	National Petrochemical Company
OMW	Osterreichische Mineralolwerke
OPEC	Organization of Petroleum Exporting Countries
OSCO	Oil Services Company of Iran
PEGUPCO	Persian Gulf Petroleum Company
SIRIP	Iranian-Italian Oil Company
SOFIRAN	A subsidiary of ERAP
SRI	Stanford Research Institute
UNCTAD	United Nations Conference on Trade and Development

HISTORICAL
DEVELOPMENTS
TO 1951

OIL AND GAS IN IRAN

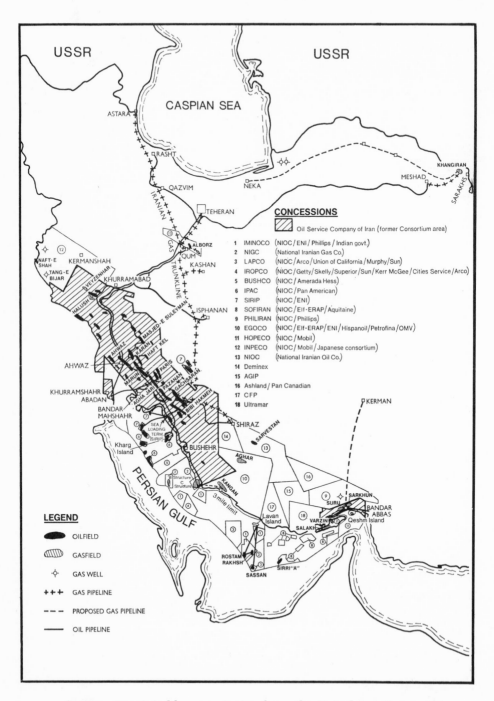

Source: Reprinted by permission from the Petroleum Economist, June 1975.

1

THE POLITICAL AND ECONOMIC IMPLICATIONS OF OIL CONCESSIONS

The purpose of this chapter is to provide an introduction to the Iranian oil industry in the first half of the twentieth century. The chapter is divided into two sections: first, a review of the D'Arcy concession and its political and economic implications, and second, an investigation of the role of oil industry in development of the economy.

BACKGROUND TO THE CONCESSION

Despite its glorious history, Persia was a weak and underdeveloped country in the nineteenth and early twentieth century. The growing power of the tzarist regime to the north of the country led to some wars between Russia and Persia which eventually destroyed the Persian military power and paved the way for ever-increasing Russian influence in the country. Russia's relentless search for an outlet to the open sea was the main cause of her interest in Persia.

Political rivalry between Russia and Britain, however, started with a fear of the Russian advance toward India. This fear placed Persia in a prominent position in British strategic planning. It is to be emphasized that during the nineteenth century, British interest in Persia was primarily strategic, commercial considerations occupying a secondary place. British policy in Persia in the nineteenth and twentieth centuries can be considered from four points of view: the first embraced the commercial and strategic interests of the British as a maritime power in the possession of India; the second, sharp rivalry between Britain and Russia in political and economic affairs; the third, cooperation with Russia for the maintenance of peace on Indian borders; and finally, the partition of Persia along the line of the development of the oil industry and the expansion of the British influence in political life of the country.[1]

There was a time when the British statesmen were considering a policy that sought to strengthen Persia: they wanted Persia to stop the tide of Russian movement toward India, but later the British aimed at weakening Persia. They followed a policy of demanding equal privileges and concessions in the south to balance every Russian gain in the north. This policy stripped Persia of her sovereignty and culminated in the Anglo-Russian Treaty of 1907, which effectively partitioned Persia by dividing it into two spheres of influence with a neutral zone in between. Both countries agreed to respect Persian sovereignty and independence. Neither party in fact did so.

The discovery of oil in Persia was a turning point in British policy toward Persia. For the first time strategic considerations gave way to commercial interests. As we shall see in the course of this chapter, the British stripped the central government of its powers by making separate deals with the rebellious tribes.

In March 1915 the second Anglo-Russian convention, referred to as "the Constantinople Agreement," was signed. The agreement gave to the British the oil-rich neutral zone in return for Constantinople and Eastern Turkey. At the same time the Russians (with the approval of the British) were advancing their troops toward Tehran with the final aim of a second and complete partition of Persia. Fortunately, the 1917 October Revolution in Russia took place before the final annexation was near completion, and Lenin ordered all the Russian troops to evacuate Persia. Had it not been for this single event, Persian oil would perhaps have caused the country to become colonies of Russia and Britain.

The population of the country in 1900 was just under 10 million and the rural population constituted over 80 percent of the total.[2] The Persian society contained various groups, none of which was in a position to exert any influence on the monarch. The absentee landlords treated the peasants as their slaves, extracting what they produced and leaving them at bare subsistence levels. The religious hierarchy was strong but aligned to the landowners who paid for their maintenance. The bazaar or merchant class, unlike the Western middle-class bourgeoisie who initiated so many reforms, never developed any political consciousness capable of transforming the Persian society. Nor was there in Persia a well-educated aristocracy of sufficient coherence to influence the shahs. Although some of the nobility held official positions of responsibility, most were loafers who placed a heavy burden on the limited resources of the treasury.

The shahs had three major sources of income: indirect taxes on goods such as tobacco and custom duties; sale of official posts to

those who gave them handsome presents and who in turn sold jobs to
their subordinates; and finally, borrowing from foreign banks, using
as collateral pledges of certain rights for foreigners in Iran. By the
end of the nineteenth century the rulers found such income insufficient
for their extravagant expenditures and looked more and more toward
concessions to supplement their incomes.[3]

    In short, the political and economic climate of the country at
the turn of the century was that of corruption, deceit, and decay. It
goes without saying that given this kind of political atmosphere no
proper care could have been taken in formulating the D'Arcy concession, which has overshadowed the Iranian history in the twentieth
century.

                    The D'Arcy Concession and the Origins
                         of the British Involvement

    In 1892 the French geological magazine Annales des Mines
published a report on the prospects of oil discovery in Iran. This
report was based on the investigations of a French government
scientific mission which had been sent to Iran in the early 1890s.
This report however, did not interest potential investors before 1900.
It was indeed the Director-General of the Persian Customs, Antoine
Ketabji-Khan, who first brought the matter to the attention of European
financiers, among whom was the Britisher William Knox D'Arcy.

    Apart from the favorable geological conditions in the south and
southwest of Iran, the sociopolitical environment was of considerable importance in persuading D'Arcy to enter negotiations with
Persia. D'Arcy understood the fact that, given the corrupt political
system and the impotence of the monarchy, there would be few dangers to his potential investment. On the other hand, the British
government looked upon D'Arcy's quest to obtain a concession as a
mission well deserving of support. Sir Arthur Harding, the British
Minister in Tehran, was instructed to do all he could in securing an
oil concession for D'Arcy.

    The 1901 D'Arcy concession provided for a 60-year exclusive
privilege to "search for, obtain, exploit, develop, render suitable
for trade, carry away and sell natural gas, petroleum, asphalt, and
ozokerite" throughout the whole of the empire, with the exception of
the five northern provinces (Azarbayejan, Gillan, Mazanderan,
Khorasan, and Asterbad) which were under the influence of Russia.
The concession embraced an area of 500,000 square miles--an area
larger than that of France, Great Britain, and Germany put together.
It granted D'Arcy the exclusive right to lay pipelines to the Persian
Gulf, a privilege which proved to be D'Arcy's greatest deterrent

against potential competitors. All land taxes and import duties were
waived and Article 14 of the concession bound the Persian government
to protect the physical assets and personnel of the company. But the
same article specifically denied the company the power to ask for
compensation from the government under any pretext. The concessionaire could form one or several companies to operate the concession, all the companies receiving the same privileges awarded to
D'Arcy. The concession agreement was signed in 1901.

In return D'Arcy and his companies were to pay the Persian
government, within a month of their formation, £20,000 in cash and
£20,000 in stock of the companies, plus 16 percent of their net profits
per annum ("net profit" implied net integrated profit of all companies
dealing with Persian oil). In addition, a fixed sum of 2,000 tomans
(£1,800) per annum was to be paid to the government in lieu of taxes.
There were to be two arbitrators in case of any disputes, one appointed by each party, and an umpire selected by the arbitrators
whose decision was to be final. The most important economic issue
which was embodied in the concession was the linking of royalty payment to profits. This, as we shall see later, raised innumerable
problems and difficulties. As far as D'Arcy was concerned, this
would provide him with a protection in years of little or no profit.
This method also seemed attractive to the Persian government in
their hope that the industry would be in a state of continuous pros-
perity--which in fact did not turn out to be the case. Payments under
this method were subject to fluctuations varying with those of the
companies' profits, a situation which was contrary to the Persian
government's expectation of securing a more or less steady stream
of cash to the treasury.

In 1903 the First Exploration Company (FEC) was formed with
a capital of £600,000 in shares of £1 each. G. B. Reynolds, a man
of remarkable character, was put in charge of the operations. The
first few wells, although containing oil, were not large enough to be
commercially viable. Under the terms of the concession the central
government in Tehran was responsible for protecting the companies'
operations, but weak and corrupt as it was, it lacked control of the
Bakhtiari tribesmen who had immense powers in the southern prov-
ince of Khuzestan, where Reynolds was operating. To protect the
operation from raids, D'Arcy formed the Bakhtiari Oil Company,
with a capital of £400,000 in £1 shares, and presented 3 percent of
the shares to the tribal chiefs. D'Arcy also agreed to pay the tribesmen £2,000 per annum for safeguarding the drilling equipment and
supplies, and £1,000 per annum for safeguarding the pipelines.

D'Arcy's FEC spent five years and nearly a quarter of a million
pounds sterling before oil in commercial quantities was struck in
southern Persia. During this period, D'Arcy's funds were exhausted

and his efforts to raise capital in London met with failure. He therefore decided to sell to foreign interests. During the same period,
John Fisher, who had dreamed of the substitution of oil for coal in
the British ships since 1882, was appointed the First Lord of the Admiralty. His first act was to appoint an oil committee to investigate
the feasibility of switching the naval fuel from coal to oil. In view of
the likelihood of D'Arcy selling his concession to foreign interests,
the Admiralty approached Lord Strathcona of the Burmah Oil Company and suggested that his company provide funds for the D'Arcy
venture to develop the Persian concession.

        In this way the Concessions Syndicate Limited was formed in
1905, with D'Arcy as a director and with adequate capital to continue
operations in Iran. The exploration, however, still met with failure.
Reynolds moved his operations to Masjid-i-Soleiman (then called
Maidan-i-Naftoon). In this place the well M1 was drilled for 1,000
feet. There Reynolds received a telegram from the directors of the
syndicate saying that funds were exhausted, and that the decision
reached finally and irrevocably was that he was to cease work, dismiss his staff, dismantle anything worth the cost of transporting to
the coast for reshipment, and come home. Reynolds thought of an
idea to continue work. On account of possible errors in the coding
of such an important message he said it would not be safe to rely on
the telegram. He decided to wait until official confirmation arrived.
This with luck was expected to take a month. Meanwhile the drilling
went on. A fortnight later, on May 26, 1908, M1 having reached a
depth of 1,110 feet, struck oil in large quantities. This was the
extraordinary account of the Persian oil discovery.[4]

        In 1909 the Anglo-Persian Oil Company (APOC) was established
to take over the Persian oil operations. The company's initial capital of £2 million was largely held by the Burmah Oil Company. The
transfer of the concession from D'Arcy to APOC has frequently been
used as subject by imaginative storytellers. The stories range from
the attempted assassination of D'Arcy in Cairo by British secret
agents to the continuous pressure brought to bear on him by the
British government to transfer his holdings to APOC. One of the
most widely quoted stories is about the role of Sidney Reilly, the
British secret agent, in obtaining the concession from D'Arcy.
Reilly was said to have disguised himself as a priest, and having
persuaded D'Arcy that he should set up a charitable company to con-
vert Persians to Christianity, obtained the concession from him.
An examination of the Annual Reports of the APOC in the 1909-16
period shows that the story is false. In fact D'Arcy never "disappeared" from the scene.[5] He was alive and well and on the Board of
Directors of the company for seven years. He retired at the end of
1916 because of ill health.

An important point, which is unfortunately missed in most of
the available literature on the formation of APOC is that when the
company was formed in 1909 it did not take over the operation of the
FEC and the Bakhtiari Oil Company. These two companies were
maintained as subsidiaries, concerned with the producing activities
of APOC. FEC and the Bakhtiari Oil Company were owned by APOC
to the extent of 97 percent (3 percent Bakhtiari Khans). APOC's
relationship with regard to these companies was very much like a
holding company.

The Anglo-Persian Oil Company continued its operations with
success in the south of Iran, but to expand its activities more capital
was required. The extra funds were eventually to come from the
participation of the British government in the venture. It was mainly
through the efforts of Winston Churchill, who was appointed the First
Lord of the Admiralty in 1911, that the British government decided
to buy a controlling interest in the Anglo-Persian Oil Company in
1914. The total capital of the company was raised from £2 million
to £4.2 million, with the British government investing £2.2 million
in the company. The share of the government was then around 52.5
percent, Burmah Oil Company 22.5 percent, and the public 25 percent. There was a great deal of debate in the House of Commons for
and against the British government's participation in the venture,
details of which are well documented.[6] Churchill talked of the advantages to the Admiralty and Great Britain that the purchase would
bring. It would free the navy from exclusive reliance on a world
market dominated by Standard Oil and Shell and would lay the basis
for a British entity which the navy could count on in war and peace.
He reiterated his belief that the development of the oilfields should
be undertaken "according to naval and national interests" of the
British government. His arguments showed his lack of consideration
for the Persian interests or, indeed, for the possible implications
this might have in the Persian economy. Ramsey Macdonald,
A. W. H. Ponsonby, and some other members of the House voiced
their opposition to what they saw as direct British interference and
weakening of the Persian government. But Churchill got his way by
a vote of 254 to 18. As it turned out Ponsonby and Macdonald were
right in their predictions. Britain signed an agreement with Russia
in 1915 which effectively divided Persia into two colonies.

A direct consequence of the British government's involvement
in the activities of APOC was the 40-year supply contract signed between the British Admiralty and the company. The contract was for
the sale of fuel oil to the navy at special discount prices. Although
the prices were never officially disclosed, there were agreed to be
special discounts from the market prices. The supply contract was
one of the major issues raised by the Persians in the nationalization

debates. According to Persians, the cost of oil to the Admiralty during the 40-year life of APOC ranged from 30 to 40 cents per barrel, against the market prices fluctuating between 90 cents to $2.43 per barrel.[7] According to Churchill this contract saved the British government £40 million between 1914 and 1923.[8]

1. The original government investment of £2.2 million in ordinary shares, had by 1923 become one of 5 million shares and the appreciation in value of this at current price was approximately . . . . £16 million
2. The government had received in taxes and dividends and interest . . . . £6.5 million
3. The supply contract had enabled the government departments to save on purchases, as compared with market prices . . . . £7.5 million
4. The sum saved by the competition of APOC from other companies and the saving on oil prices under the supply contract. . . . . £10 million
                                                                   £40 million

Churchill had this to say in 1923 about the benefits to the British government of involvement in APOC: "And so it all went through. Fortune rewarded the continuous and steadfast facing of these difficulties by the Board of Admiralty and brought us a prize from fairyland far beyond our brightest dreams."[9] He added that the total profits realized from this investment were not only large enough to pay for the cost of the fuel to the ships, but also for the addition of great fleets of ships to the British Navy, without costing the taxpayer a single penny.[10] Persians claimed that the benefit to the British government in the 40-year life of APOC was £500 million, and that the contract alone saved the British government 15 percent more than the total amount paid to the Persian government in taxes and royalties during the 40 years of operation of APOC in Persia.[11] Although there is no evidence to support the figures presented by the Persians, it shows the feeling of the people toward the participation of the British government in the company. It is interesting to note that the payments to the Persian government in the 1911-23 period were around £3.5 million compared to the £40 million figure quoted by Churchill. The £40 million benefit to the British government shown by Churchill is also suspect. Let us remember that this was an attempt to glorify his own actions. Even excluding the appreciation of the value of the shares, the Churchill figure stands at £24 million benefit to the British government in the 1911-23 period, while the total net profit of APOC (after payments to Persia) was just under £11 million in the same period. Considering that the dividend payments

were around 15 percent on average, and that no dividends were paid
out before 1916, we can see that there are inconsistencies in Churchill's figures.[12]

The benefits to the British government were not all economic.
Britain used APOC to expand its political dominance of the country by
bribery and force where necessary. Arthur C. Millspaugh, an American who served as an Administrator General of Finance in Persia
from 1922-27 and 1943-45, argued that despite the British government's pledge of noninterference in the commercial management of
the company, APOC became "to all intents and purposes an arm of
the British Admiralty and the British strategic policy."[13]

Armitage Smith and the "Interpretive" Agreement

Two points in the D'Arcy concession caused considerable disagreement between APOC and the Persian government--the revenue
payments by the former, and the security obligations of the latter.
The D'Arcy concession clearly specified that the Persian government
was to obtain 16 percent of the net profits of all companies formed
to exploit Persian oil. APOC questioned this right by arguing that
since these subsidiaries were operating outside Persia, they should
not be liable to pay royalties. Also the concession shows that the
Persian government was not liable for any damage caused to APOC
assets, should the government fail to protect the companies' properties (Article 14). APOC disregarded this provision and frequently
deducted from the royalties the sum paid to local tribes as a bribe.
In 1915, APOC paid only 13 percent of net profits to the government,
saying that the other 3 percent was spent on repairing the damage
caused by the raids from local tribes. It is interesting to note that
these violations of the agreement started mainly after the British
government became a shareholder in the company, and also in the
same year as the Anglo-Russian Agreement of 1915, for the second
partition of Persia, was signed--with the British troops in the south
and the Russian troops advancing toward Tehran from the north. The
Russian Revolution stopped the Russian advance and their soldiers
were gradually withdrawn. Lenin announced that all agreements with
"imperialists" were to be null and void and Persia was rid of Russia
for a while.

The withdrawal of Russian troops did not help the Persian
government. On the contrary, it enabled the British government and
APOC to dominate Persia completely, though no formal colonization
took place. In 1917, the British government was subsidizing not only
the Persian administration but also individual rulers, such as the Shah
(at £120,000 per year), the Prime Minister, Vossough-ed-Dowleh,

the Minister of Foreign Affairs, and the Finance Minister.[14] Under
these circumstances, APOC asked for what it called "an interpretive
agreement," to clarify the controversial points of the original D'Arcy
concession. In 1919 an agreement was signed to the effect that the
British government would lend advisors to the Persian government,
at the latter's expense, with "adequate" powers to resolve all disputes. The treaty was considered by a British official to have "turned
Persia into a private mandate for Great Britain."[15] Later the Persian government was persuaded to appoint a British Treasury Official, Sir Sidney Armitage Smith, as Persia's representative in the
negotiations.

    In December 1920 the so-called "Interpretive Agreement" was
signed. The agreement differed extensively from the original D'Arcy
concession to the detriment of the host country. It accepted the
Persian contention that the profit-sharing principle applied to all the
companies formed by APOC, but provided for large deductions from
the net profits of the subsidiary companies which were involved with
refining, transportation, and distribution of Persian oil outside Persia. In practice, with the great flexibility in accounting procedure
which an integrated company such as APOC enjoyed, the company
was able to allocate the largest portion of its profits to downstream
activities and the balance to the producing end of the operations.
In this way the Armitage Smith Agreement limited the profit-sharing
of the Persian government in the concession, to the producing activities of the companies within Persia. In return for this agreement
APOC dropped its claim for damages and paid the Persian Government £1 million in settlement of all past claims. This settled the
issue only temporarily.

    Three months after signing the Armitage Smith agreement a
bloodless revolution took place in Persia. Colonel Reza Khan marched
at the head of a Cossack detachment to Tehran and occupied the capital. As Minister of War and later as Prime Minister, he brought
together all the local rulers under one flag, by a combination of per-
suasion and force. In 1925 the Majlis (Parliament) deposed Ahmad
Shah, the last King of the Qajar Dynasty, and Reza Shah Pahlavi (the
father of the present Shah) was placed on the throne. Reza Shah was
a man of remarkable character. He was determined to put the country on the road to progress after a hundred years of weakness, corruption, and decay. His methods were often harsh but effective.
Bribery and corruption were stamped out ruthlessly, taxes were re-
formed, and universal conscription was introduced. Also, educa-
tional institutions were greatly expanded and a great deal of road and
railway construction was undertaken. Shortly after Reza Shah came
to power, he asked Lumely and Lumely, the London solicitors who
had been occasionally assisting the Persian government in legal

matters, to examine the Armitage Smith agreement. Lumely and
Lumely expressed the opinion that the agreement constituted not an
interpretation of the concession agreement, but a modification of it,
and hence required ratification by the Persian Parliament to make it
binding. They also expressed the opinion that Armitage Smith had
exceeded the powers conferred on him by the Persian government in
negotiating a settlement on such a broad basis.

The Armitage Smith agreement was never ratified, nor was it
properly repudiated. Year after year the Persian government challenged the basis on which the royalties were being calculated, but no
response was forthcoming from the British side. Feelings of frus-
tration were running high among the people and the Shah. The British
politicians did not fully appreciate the significance of the changes
which were taking place in Persia under the leadership of Reza Shah,
and continued to ignore Persian demands.

While the feelings were high in Persia, the international financial crisis of the 1930s took place. Whereas APOC paid the Persian
government £1.28 million in 1930, payments declined to £0.306 million in 1931. The decline in the royalty payments reflected the world
economic depression and particularly the depressed condition of the
oil industry. With a decline of 3.5 percent in the Persian production,
payments to Persia declined by 76 percent. During the same period
the profits of APOC declined by around 36 percent. During the
1911-32 period the Persian government received a total of £12.8
million in royalties, while the net profits of the company were £51.5
million and the U.K. taxes were £7.7 million.

The sharp decline in royalties while Reza Shah was carrying
out large development projects was the last straw for the Shah. He
warned APOC of the possibility of the cancellation of the concession,
but APOC and the British government refused to recognize the Shah's
right to cancel the concession. On November 27, 1932, the Minister
of Finance delivered formal notice of cancellation of the concession
to the resident director of APOC. The British government, however,
refused to accept Persia's right to cancel the concession and warned
of serious consequences such an action might entail for Iran. The
arguments dragged on and finally the British took the Persian government before the Council of the League of Nations in 1932.[16] The
arguments before the Council indicated the sharp differences between
the two parties. They illuminated not only the immediate issue, which
precipitated the dispute, but also the basic grievances that the concession had created. Before the Council announced a decision a
compromise was reached. The calculation of royalties was to be
made on a completely new basis.

The 1933 Concession and the Principle
of Tonnage Royalty

It is still a matter of personal opinion as to who was the winner
in the contest. Even the experts cannot agree as to which side obtained the best deal and which side capitulated first. Ford argues
that the 1933 concession was a victory for Iran,[17] while the Iranian
government claimed in 1951 that this was a setback for the country.[18]
Fatemi argues that the motivation of the company, by paying low
royalties in 1931, was to force the cancellation of the D'Arcy concession with a view to securing an extension in the period of the
lease.[19] This contention is hardly tenable. A careful examination
of the D'Arcy concession gives the answer. The D'Arcy concession
and the Armitage Smith agreement rendered the Persian government's
revenues more sensitive to a decline in APOC's income, to the extent
that the government was to share (at the rate of 16 percent) in the
profits of some APOC affiliates after certain fixed deductions were
made. Thus, during a depression year such as 1931, Persia was
left with a very small royalty, mainly as a result of the decline in
APOC's net profits. However, it is important to emphasize that the
decline in the royalties was more than proportional to the decline in
APOC's net profits. This was because the Persian government's
share of net profits at 16 percent was calculated under the Armitage
Smith agreement.

To remedy this situation, the principle of tonnage royalty was
introduced. This system was to link the royalty payments to the volume of oil production with a built-in protection device against change
in the value of sterling. APOC agreed to pay the Persian government
the sum of 4 gold shillings per ton of Persian oil extracted, and in
lieu of taxes to pay ninepence per ton (gold) for the first 6 million
tons of oil produced in any one year, for the first 15 years, and
sixpence per ton for production of oil in excess of 6 million tons
(also for the first 15 years). After the first 15 years and up to 30
years, those payments in lieu of taxes were to be raised to 1 gold
shilling for the first 6 million tons and 9 gold pence for the production in excess of 6 million tons. APOC also agreed to pay the Persian government 20 percent of the distribution of its earnings in excess of £671,250. It guaranteed that the payments to Persia from
royalties, taxes, and a percentage of distributed earnings would not
be less than £975,000 in the first 15 years or less than £10,050,000
in the next 15 years. The area under exploration was to be reduced
by 80 percent to 100,000 square miles, and the period of the concession was to be 60 years. Payments of royalties and taxes for the

second 30-year period (after 1963) were to be negotiable. The company also agreed to pay £1 million to Persia in May 1933 in settlement of all past claims.

The revised agreement required APOC to develop the Naft-o-
Shah oilfield on the Iran-Iraq border, west of Kermanshah, and to
process its output for Iran's domestic consumption. In addition, the
companies agreed to make efforts to increase their Persian staff and
place some Persian nationals in technical and commercial positions.
The provision for arbitration was to be as in the original D'Arcy concession, but the umpire was to be selected by the President of the
Permanent Court of International Justice, if the two parties could not
agree on a mutual selection. The concession contained several new
provisions to protect Persia's interests. It gave Persia the right to
appoint a "Delegate of the Imperial Government" at a salary of £2,000
per annum, payable by the company, who should have access to all
information that the stockholders were entitled to.[20] The concession
also contained provisions designed to ensure lower prices of petroleum
products for domestic consumption.[21] Finally, the company pledged
itself to noninterference in Persia's domestic life. In return the government exempted the company from all taxes except those stipulated
in the concession, and agreed not to annul or alter the concession
"either by general or special legislation in the future, or by administrative measures, or any other act of the executive authorities."
In 1935 Reza Shah decided that the official name of the country should
be changed from Persia to Iran (as it was called by the nationals).
In compliance with this order, the company changed its name to the
Anglo-Iranian Oil Company (AIOC).

In terms of profitability per ton of oil produced, the Persian
government was slightly better off during the period of the 1933
Agreement, as Table 1.1 shows. The table clearly shows the downward movement of oil payments toward the end of the D'Arcy concession and the improvement after the 1933 agreement. It is interesting to note that in most of the years of the D'Arcy concession,
the net profit per ton of AIOC rose or declined proportionately more
than the Persian government's receipts. At the same time, AIOC's
tonnage profits did not rise as fast as those of the Persian government until the end of World War II.

IMPACT ON THE IRANIAN ECONOMY

The political and economic aspects of the oil concessions in the
1901-51 period have been discussed in the preceding pages. The
present section will deal with the impact of the oil industry on the
Iranian economy.

TABLE 1.1

Per Ton Revenues of the Persian Government
and AIOC Profits for Selected Years
(pounds sterling)

Year	Persian Government Revenues per Ton	AIOC Revenues (Profits) per Ton
1913-14	0.04	0.10
1919-20	0.34	1.34
1925-26	0.23	0.96
1930	0.22	0.64
1931	0.05	0.42
1933	0.26	0.37
1935	0.30	0.47
1940	0.47	0.33
1945	0.33	0.34
1946	0.37	0.50
1950	0.50	1.06

Source: F. Fesharaki, "The Development of the Iranian Oil Industry 1901-71," Ph.D. dissertation, Surrey University, 1974, Chapter 3.

The development process of a developing economy can be studied by (a) comparing development theories with the actual situation and (b) constructing a macroeconomic model and an examination of the input-output matrices. In the case of Iran for the period under study, no detailed information was available, nor was there any input-output matrix constructed.[22] Thus one is left with the first alternative--to compare the development theories, such as balanced and unbalanced growth theories, stage theories, typology theories, and enclave theories with the actual situation in Iran. In the final analysis one is likely to find that one or a variation of a number of theories will approximately fit the Iranian situation at one point in time, while at another point another theory will be more appropriate.[23] In the following, however, we will not go through the development literature which is widely available,[24] but will rather focus on the information available to see the possible contribution of the oil industry to the Iranian economy.

The significance of a leading sector in an economy can be examined through revenue earning capacity for government spending

(indirect effects) and the interrelationship of the sector itself with the
rest of the economy (direct effects).

Indirect Effects

The oil revenues received by the Iranian government during the
period prior to Reza Shah's reign were mainly used for supplementing
the private purses of the rulers. Thereafter, these revenues were
incorporated in a general budget which was responsible for the running of the government departments and defense as well as occasional
development projects.

To understand the significance of these oil revenues, they may
be viewed in two different contexts: first, in comparison to what the
non-Iranians (AIOC, British government, and the British shareholders) had received, and second in comparison with the size of general
budget, development expenditure, and total foreign exchange receipts.

Table 1.2 illustrates the overall picture of AIOC's financial
performance, and provides an insight into the relationships between
AIOC's net profits, U.K. taxes, royalty payments, and dividends of
the ordinary shareholders.

The company profits rose from £27,000 in 1913-14 to a peak of
£4.8 million in 1926-27, but fell to £2.4 million in the depression
year of 1931. Royalties to the Persian government started from a
low of £10,000 in 1913-14, rose to £1.4 million in 1929, and fell
drastically to only £307,000 in 1931. The shareholders of the company
were particularly well remunerated during the 1933 agreement. Dividend payments rose from a high of 5 percent in the years of the Great
Depression, to 20 percent in 1936-38. Scrip bonuses (or stock dividends) of 50 percent were distributed in 1936. In the 1939-40 period
dividends fell to 5 percent, owing to a reduction in oil exports after
the outbreak of war. Later they rose to 30 percent in the years
1946-50. Had it not been for the British government's dividend limitation policy in the late 1940s, AIOC would, in fact, have distributed
larger sums.[25]

Despite the nationalization of oil by Iran in 1951 and the virtual
standstill of Iranian oil production, dividends continued at the rate of
30 percent for 1951 and rose to 35 percent in 1952, to 42.5 percent
in 1953, and to 15 percent plus 400 percent scrip bonus in 1954. The
large financial reserves accumulated by the company (which would
have raised the Iranian share of profits had it been declared as dis-
tributable profit), made it possible for the company to follow such a
generous dividend policy. Table 1.2 also shows that within 20 years,
the net profits of AIOC rose from £2.4 million in 1931 to £33.8 million in 1950--a rise of over 14 times. Payments to Iran rose 12
times during the same period.

TABLE 1.2

Production, Net Profits of APOC/AIOC, U.K. Taxes, Iranian Royalties,
and Dividends to Ordinary Shareholders in Selected Years

Year	Net Production (thousands of metric tons)	Net Profits[a] (thousands of pounds sterling)	U.K. Taxes[b] (thousands of pounds sterling)	Iranian Royalties[c] (thousands of pounds sterling)	Ordinary Share Dividends, Percent (1 pound sterling per share at par)
1913-14	278	27	--	10	--
1917-18	912	780	--	330	10
1921-22	2,365	3,779	--	593	20
1926-27	4,909	4,800	--	1,400	12.5
1929	5,545	4,274	--	1,437	20
1931	5,843	2,413	--	307[d]	5
1936	8,330	6,123	911	2,580	25 plus 50% scrip bonus
1940	8,753	2,842	2,975	4,000	5
1944	13,487	5,677	10,636	4,464	20
1948	25,270	24,065	28,310	9,172	30
1950	32,260	33,103	50,707	16,032	30
1951	16,886	24,843	27,374	8,326	30

[a]After taxes and royalties. Consolidated accounts were not prepared before 1958. Includes profits of AIOC's subsidiaries not operating in Iran.

[b]Includes company income tax and corporate profit tax. From 1911-32, U.K. taxes amounted to 7.7 million pounds sterling.

[c]In 1940 AIOC paid an indemnity of 1.5 million pounds sterling to the Iranian government and guaranteed £4 million for the 1940-44 period. This was to compensate for the company's inability to export sufficient crude and products during the war years.

[d]After the 1933 agreement this figure was revised to £1.3 million.

Sources: L. Nahai and C. Kimble, "The Petroleum Industry of Iran," U.S. Department of Interior, Bureau of Mines, 1963, p. 17; Z. Mikdashi, A Financial Analysis of Middle Eastern Oil Concessions, 1901-65 (New York: Praeger Publishers, 1966), pp. 45-46; and National Iranian Oil Company.

A word of caution with respect to the interpretation of Table 1.2. It must be noted that the net profits shown above are on the whole of the integrated network of AIOC and not only on its producing subsidiaries in Iran. U.K. taxes are also levied on all the operational profits. Although AIOC's source of crude for most of this period was Iran, one should nevertheless guard against making such comparisons without qualification, particularly as AIOC's accounts were not always consolidated.

Let us now consider the size of the oil revenues in relation to the budgetary receipts and expenditures. Until 1949 Iran had only one budget which included development expenditures. Since no separate development budget was prepared, it is difficult to accurately measure the contribution of the oil revenues to the development expenditure. However, some scanty information is available. Up to 1927 almost all of the receipts were earmarked for the current expenditure, but during the 1930s development expenditure became more significant, accounting for between 30 to 40 percent of the total government expenditure.[26] The increase in development expenditure was a direct result of Reza Shah's initiatives. In 1929 the number of primary schools was about 1,000, compared to 50 a decade earlier. High schools and institutes of higher education were similarly developed, while for the first time groups of best students were sent at state's expense to study various branches of science and technology in foreign countries. Health services were expanded and hospitals built, but perhaps the most important activity was that of road construction and the trans-Iranian railway system which laid the basis of the infrastructure necessary for the development of the country. Defense expenditures during this period were also significant, amounting to 40 percent in the 1920s and around 23 percent of the total budget in the period 1937-49. These expenditures were necessary at the time for the introduction of national conscription and the need to unify the divided country.

Reza Shah's attempts to focus on development planning were disrupted by World War II. The public development outlays fell to 19 percent in 1942, 7 percent in 1945, and 15 percent in 1949. (See Table 1.3.)

As we can see from Table 1.3, the oil revenues have at no time exceeded 15 percent of the total revenues. Indeed, the largest single source of income during this period was custom duties, while taxes accounted for over two-thirds of total revenue.

One important danger of this type of comparison is that one may mistakenly assume that local and foreign sources of income have the same significance. While the above type of comparison is useful in identifying the trends and the relative orders of magnitude, it should be noted that foreign exchange is basically different in nature to local currency. Foreign exchange can only be used to buy imports and cannot be spent domestically. Thus, a more useful assessment of the oil revenues would require its comparison with the total foreign exchange receipts of Iran. Unfortunately, adequate data for such purpose are not available until the mid-1940s. In 1947-50 the oil payments, including the purchases of local currency by AIOC, accounted for 66 percent of Iran's $154 million average annual foreign exchange earnings.[27]

TABLE 1.3

Estimates of the Iranian Budget by Major Components;
Sources and Uses of Funds, 1937-49
(millions of rials)

	1937	1942	1945	1947	1949
Sources					
Income tax	130	198	644	510	900
Excise tax	180	175	360	320	550
Custom duties	442	307	457	1,312	1,679
Other taxes[a]	412	1,259	1,445	1,674	2,326
Total taxes	1,164	1,939	2,906	3,816	5,455
Public enterprises[b]	118	349	798	739	1,047
Oil payments	206	347	512	677	901
Other revenue	156	109	196	327	382
Total budgetary receipts	1,644	2,744	4,412	5,559	7,785
Uses					
Development expenditure	662	512	313	991	1,638
Other expenditure[c]	976	2,251	4,099	7,130	9,479
Total expenditure	1,638	2,763	4,412	8,121	11,117
Contribution of oil revenues to total expenditure (percent)					
	13	13	12	8	8

[a]Includes inheritance tax, land tax, and fiscal monopolies.

[b]Includes post and telegraph, government domains, public entities, and investment income.

[c]Includes defense, administration, health and education, interest payments, and debt redemption.

Note: The gap between income and expenditures after 1945 was presumably filled by domestic and foreign borrowing and the like.

Sources: United Nations, Public Finance Information Papers (Iran); and J. Amuzegar and A. Fekrat, Iran: Economic Development Under Dualistic Conditions (Chicago: University of Chicago Press, 1971), Part I.

Direct Influences

The direct impact of the oil industry on the Iranian economy
may be considered through forward and backward linkages--that is,
the flow of resources to and from the domestic economy. This will
provide an insight into the degree of integration of the leading sector
with the national economy and the extent by which this sector stimu-
lated growth in other sectors.

Let us first look at the forward linkages--the flow of resources
from the oil sector to the rest of the economy. Forward linkages
could comprise either AIOC's participation in local industries through
investment in by-product ventures linked to oil production and refining as well as diversification into nonoil activities, or the sale of oil
products in Iran.

There is no evidence to suggest that AIOC was involved in any
local venture in Iran. Indeed, the company's annual reports and
Chairman's reports show that such matters were never seriously
considered by the management. The company had simply no other
interest in Iran, except for those in its concession agreement.

The only significant forward linkage was the consumption of
oil products in Iran during this period. There was no real effort on
the part of AIOC to encourage the use of oil for home cooking, heating, and other uses. Indeed, the oil products consumed by the
Iranians did not all come from AIOC. There were three major sources
of supply: the AIOC's own sales of its domestically refined products,
AIOC's imports from abroad, and imports from Russia. Table 2.4[28]
shows the relationship between Iranian production, sales, and exports. It can be seen that until the late 1920s more than three-quarters
of the oil consumption was imported. In the late 1930s and 1940s
the imports fell drastically, and almost all of the oil products were
supplied by the domestic industry. Until 1937, the Soviet Union's
exports to Iran accounted for the bulk of Iranian oil consumption.
The lack of integration between the domestic economy and the oil
industry is self-evident through the data in Chapter 2.

Backward linkages or the flow of resources from the national
economy to the leading sector could take the form of

(a) participation of the Iranian nationals in providing equity
capital
(b) providing local materials for use by the company and its
employees
(c) providing work at manual, technical, and managerial levels.

Let us consider each of these linkages in turn.

(a) AIOC had decided from the beginning that the company should remain exclusively in British hands. The company did not offer any shares for sale on the domestic market, although any Persian would have presumably been able to buy them on the London Stock Exchange. In 1919 the Persian Foreign Minister solicited the British Foreign Secretary in London for the right to purchase a few shares out of a new issue of £7.5 million to be made in 1920. His request was turned down.[29] The APOC management even went so far as to persuade the Bakhtiari Chiefs, who held shares to the value of £15,540 in FEC and over £11,000 in the Bakhtiari Oil Company in 1911, to sell all theirs. By 1929 they had sold all their equity capital, despite the fact that the company was increasingly prosperous.

(b) As far as local material production was concerned, APOC offered little or no stimulus. Most of the requirements of the company were imported from abroad, including food and drink. Contrary to Article 7 of the concession which permitted the import free of duty only for items necessary for the operation of the company, frequent abuses of the customs privileges were reported. The British Minister in Tehran once complained to the British Foreign Office of the unreasonable attitude of the company and asked for a ruling against Greenway (Managing Director of APOC) on the point of free duty imports.[30] Indeed, the Iranians alleged that most of the household goods of the company and part of the industrial supplies such as cement could have been produced competitively in Iran.[31] However, no ruling was issued from London and within a few years all the oil companies in the Middle East adopted the same position.

(c) So far as the employment opportunities were concerned, the company's operations had only marginal effects on the overall employment. With regard to senior or managerial jobs, the opportunities were very small indeed. But the effect was rather more significant on the unskilled labor force. According to International Labor Office (ILO) figures, in 1949 less than 1 percent of the Iranian work force was employed by the company (AIOC). Only 9 percent of these employees were among salaried staff--that is, senior employees. The rest were wage earners or unskilled workers. Among the salaried employees, the number of "graded" (high-ranking) Iranians was about one-third of the British staff. On the other hand, the ratio of "nongraded" employees was five Iranians to one foreigner, most of them Indians. There were no Iranians assigned to top managerial positions within the company.[32]

Table 1.4 shows the employment situation in Iran during this period.

TABLE 1.4

Employment in the Iranian Oil Industry, 1939–51

Year	Iranians	Non-Iranians	Not Specified	Total
1939	15,060	2,723	n.a.	17,783
1940	13,380	2,273	n.a.	15,653
1941	10,980	2,079	n.a.	13,065
1942	11,654	1,803	n.a.	13,457
1943	16,389	2,864	n.a.	19,253
1944	16,485	3,380	n.a.	19,865
1945	21,781	4,030	12,143	37,954
1946	24,889	4,520	12,461	41,870
1947	28,221	4,228	11,065	43,514
1948	29,917	4,306	12,189	46,412
1949	32,011	4,477	16,410	52,898
1950	31,875	4,500	n.a.	n.a.
1951	50,662	4,271	12,951	67,884

Note: The figures for 1939–44 are small due to incomplete information on contract labor. Contract laborers were presumably all Iranian.

Figures for 1950 are taken from M. Nemazee and S. Nakazian, "Iran Presents Her Case for Nationalization," Oil Forum, 1952.

Source: L. Nahai and C. Kimble, "The Petroleum Industry of Iran," U.S. Department of Interior, Bureau of Mines, 1963, p. 20.

SUMMARY AND CONCLUSION

The D'Arcy concession which has overshadowed the Iranian history in the twentieth century was awarded by corrupt and inexperienced rulers who were dependent on foreign support for their survival. APOC/AIOC with the backing of the British government continued to abuse its privileges and sought to modify the original concession in its own favor. The major disputes between Iran and the concessionaire were the failure of the company to increase its Iranian employees and its dividend restriction policy which allowed for accumulation of large financial reserves used for diversification of its sources supply in Kuwait and Iraq--areas in competition with Iran for crude production. Moreover, the financial arrangements had given the British government an important power over the Iranian oil revenues. By imposing anti-inflationary measures such as dividend restriction policies, or increasing corporation taxes at home, the British government had on many occasions reduced the royalties payable to Iran. Net profits of the company were arrived at after U.K. taxes had been deducted, and the dividends payable to Iran followed the same rules laid down by the British for the ordinary shareholders.

The development impact on the Iranian economy was also negligible. Iran's oil revenues were small compared to the domestic budget and development expenditure, as well as U.K. taxes and APOC profits. There was practically no direct effect, through forward and backward linkages, on the Iranian economy. The oil industry was a foreign-oriented entity superimposed on an agrarian structure.

Perhaps the most fundamental source of Anglo-Iranian misunderstanding was not purely economic. There was an important political factor involved concerning the participation of the British government in the concession and their unquestionable support of the company. The Iranian attitude had considerably changed during the reign of Reza Shah. The Shah had created a sense of national unity and pride among the masses after a hundred years of rule by corrupt and degenerate rulers. Unfortunately, the British chose to ignore the massive tide of nationalism which was swaying against them and continued their policy and interference and gunboat diplomacy--the kind of policy which brought about the nationalization of Iranian oil in 1951.

NOTES

1. For a detailed analysis, see N. S. Fatemi, A Diplomatic History of Persia (New York: Whittier Books, 1952).

2. J. Bharier, "A Note on the Population of Iran 1900-66," Population Studies 22, no. 2, July 1968.

3. For a list of other concessions, see B. Mostofi, "A Review of the History of Oil Development in Iran," Bulletin of the Iranian Institute of Petroleum (Tehran), November 1961, pp. 104-12. See also S. H. Longrigg, Oil in the Middle East (London: Oxford University Press, 1968).

4. Mostofi, op. cit.

5. For details of these stories, see N. S. Fatemi, Oil Diplomacy (New York: Whittier Books, 1954), p. 8.

6. Parliamentary Debates, House of Commons, Vol. 63, June 1914.

7. M. Nemazee and S. Nakazian, "Iran Presents Her Case for Nationalization," Oil Forum, 1952.

8. W. Churchill, The World Crisis, 1911-1914 (New York: Charles Scribner's Sons, 1923), p. 134.

9. Ibid., p. 132.

10. Ibid., p. 140.

11. Nemazee and Nakazian, op. cit.

12. U.K. taxes for 1911-32 were 7.7 million pounds sterling, Even with the addition of entire U.K. taxes to the net profits of the company, still Churchill's figures seem to be exaggerated. For details, see F. Fesharaki, "The Development of the Iranian Oil Industry 1901-71," Ph.D. dissertation, Surrey University, 1974, Chapter 3.

13. A. C. Millspaugh, Americans in Persia (Washington, D.C.: Brookings Institution, 1946), p. 162.

14. Fatemi, Oil Diplomacy, op. cit., chap. 8.

15. Documents on British Foreign Policy 1919-32, quoted in Z. Mikdashi, A Financial Analysis of Middle Eastern Oil Concessions, 1901-65 (New York: Praeger Publishers, 1966), p. 16.

16. For details, see G. W. Stocking, Middle East Oil (London: Penguin Press, 1971).

17. A. W. Ford, The Anglo-Iranian Dispute of 1951-52 (Berkeley, Calif.: University of California Press, 1954), chap. 2.

18. Nemazee and Nakazian, op. cit., pp. 83-86. See also M. Ghassemzadeh, An Analytical Comparison of 1933 and 1954 Contracts (Tehran, 1968). In Persian.

19. Fatemi, Oil Diplomacy, op. cit., p. 159.

20. For the text of the agreement, see C. Hurewitz, Diplomacy in Near and Middle East--A Documentary Record, 1914-1956 (Princeton, N.J.: D. Van Nostrand Company, 1956), pp. 188-96.

21. The Iranian customers were to receive a flat 10 percent discount on all products, while the government received a 25 percent discount (see Chapter 2).

22. The first input-output table in Iran was constructed in the mid-1960s.

23. For a pioneering work, see J. Amuzegar and A. Fekrat, Iran: Economic Development Under Dualistic Conditions (Chicago: University of Chicago Press, 1971). See also R. Miksell, ed., Foreign Investment in the Petroleum and Mineral Industries (Baltimore: Johns Hopkins Press, 1971); Fesharaki, op. cit., chaps. 3 and 7.

24. For a good bibliography of the relevant development literature, see G. M. Meier, ed., Leading Issues in Economic Development (London: Oxford University Press, 1970).

25. Chairman's Report at the 40th Ordinary General Meeting of AIOC, 1949.

26. United Nations Public Finance Information Papers: Iran, ST/ECA/Ser.A/14, New York, 1951, p. 2.

27. Amuzegar and Fekrat, op. cit., p. 23.

28. See Chapter 2 for details.

29. E. L. Woodward and R. Butler, eds., Documents on British Foreign Policy 1919-1939, First series, Vol. IV (London: Foreign Office, 1952), p. 1258.

30. League of Nations, February 1933, cited in Mikdashi, op. cit., p. 39.

31. Nemazee and Nakazian, op. cit., p. 93.

32. International Labour Office, Labour Conditions in the Oil Industry in Iran (Geneva, 1950), chaps. 2 and 3. Note: In 1968 the active labor force in Iran was 7.8 million, of which 7.2 million were employed. In the same year just over 41,400 were employed by the oil industry, or 0.5 percent of the total labor force. In the light of these data, ILO's estimate of 7 million labor force is exaggerated, and in fact the percentage of work force employed by the oil industry was much smaller.

CHAPTER

2

THE SUPPLY AND DEMAND FOR OIL IN THE DOMESTIC MARKET

This chapter describes the Iranian oil industry in the prenationalization period, and in particular the development of the distribution network and consumption. Because the data are often inadequate, the treatment of some aspects of the industry is necessarily brief.

PRODUCTION AND REFINING

The first oilfield to be discovered in Iran was the Masjid-e-Soleiman oilfield in 1908. From then onward the company struck oil in many parts of the country, making Iran the largest Middle Eastern oil producer until 1951. Table 2.1 shows the development of the oilfields and the rate of production in this period. Agha-Jari was the largest producer in 1950, followed by Haftgel. The total production of Iranian oilfields in 1950 was 659,000 barrels per day.

The first Iranian refinery to be built was at Abadan. After the discovery of the Masjid-e-Soleiman oilfield, a pipeline with a capacity of 2.9 million barrels was laid to Abadan. The construction of the Abadan refinery started in 1908 and was completed in 1911--concurrently with the start of commercial production in Masjid-e-Soleiman. As soon as other fields were discovered they were connected in Abadan by various pipelines. In 1950 the Abadan refinery, then the largest in the world, reached its peak capacity of 500,000 barrels per day.[1]

In the pre-World War II period, oil exports were mostly in the shape of oil products and the refinery process took place near the producing areas. Some European countries--notably France, Germany, and Italy--had local refineries which provided much of their domestic requirements. Other European countries, and in particular Britain, had little refining capacity. After the war, the foreign

26

exchange shortage made it difficult for the latter group of countries
to pay for the value added by refining the products abroad. The continued pressure of the consuming nations in Western Europe and other
countries persuaded the international oil companies to build refineries near the consumption centers.

TABLE 2.1

Chronology of the Development of Iranian Oilfields

Oilfield	Date of Discovery[b]	Date of First Commercial Exploitation[b]	Production in 1950[c] (thousands of b/d)	Rank as Producer in 1950
Masjid-e-Soleiman	1908	1911	61.0	3
Naft-e-Shah	1923	1936	n.a.[d]	--
Haftgel	1928	1928	193.0	2
Ghachsaran	1928	1940	42.0[e]	4
Naft-e-Sefid	1935	1945	25.0	5
Agha-Jari	1937	1945	323.0	1
Pazanan[a]	1937	1943	--	--
Lali	1938	1948	15.0	6

[a]Pazanan was not considered an oilfield until 1960. It produced
natural gas during the World War II period.
   [b]From L. Nahai and C. Kimble, "The Petroleum Industry of
Iran," U.S. Department of Interior, Bureau of Mines, 1963, p. 31;
and Statistical Yearbook of Iran (Tehran, 1971), p. 194.
   [c]From Handbook of Iranian Oil Operating Companies, 1963,
p. 12.
   [d]N.a.--data relating to Naft-e-Shah oilfield are not provided
in the Consortium reports because the oilfield was taken over by
NIOC in 1954.
   [e]Refers to production in 1954.

     The change in international refining policy of all oil companies,
and in particular AIOC, diminished the international role of the
Abadan refinery. Although Abadan is still one of the largest refineries in the world, its role in satisfying the domestic demand in Iran
has been greatly reduced.[2] The Kermanshah and Naft-e-Shah

refineries were built in 1935 by AIOC, mainly to cater for the domestic demand. In 1950 their capacity was around 5,000 b/d.[3] The Masjid-e-Soleiman topping plant was constructed in 1927 by AIOC. The topping plant produced fuel oil for domestic use and export. The capacity of the topping plant was around 22,000 b/d in 1950.[4]

MARKETING ORGANIZATION AND DISTRIBUTION

After the expansion of the oil industry in the late nineteenth century, American companies such as Socony and British companies such as Burmah Oil, which were already active in the oil business in the Middle and Far East, started selling petroleum products in Iran. The products were imported in metal drums through the ports in the Persian Gulf and marketed under various brand names, such as "Tiger Brand," "Victoria Brand," and "Elephant Brand." The Soviet oil agency "Persaznaft" also imported kerosene in metal drums, each holding a maximum of 80 liters, and marketed it in northern Iran. The most popular oil product--kerosene--was in this way getting to the south and north of Iran, and over time its rate of consumption increased noticeably, its use being no longer confined to lighting, but extended now to cooking as well.

Prior to 1930 there existed keen competition between the Soviet agency and APOC for the sale of oil products in the southern and central districts of Iran. APOC is said to have resorted to changing its prices up to three or four times a day and employing street vendors with packhorses.[5] In 1928, under pressure from the Iranian government, APOC agreed to set up a centralized distribution department in Tehran.[6] Later, the terms of the 1933 agreement gave the company monopolistic rights to supply the entire domestic requirement of the country--rights which it did not fully exercise until the late 1930s.

Let us at this stage say something about prices. Although the bulk of imports were dominated by Russia for most of the period, AIOC accounted for the greater part of the remaining imports as well as all local production. AIOC sold its products at prices fixed on the basis of those prevailing in the Roumanian port of Constanta. It is very hard to determine the actual prices prevailing in Iran over the period under study, because AIOC sold its products through retailers who resold at various profit margins. There were no government or company controls on prices, and these prices depended on two major factors--the transport cost component of the prices and the profit margin required by the retailer.

The Russians, however, enjoyed partial exemption from customs duty payments until 1928, and although AIOC/Persaznaft

competition did not have a purely commercial motivation behind it,
nevertheless the partial exemption from customs provided an advantage in terms of competition for the Russians. In 1928 Reza Shah
canceled this special privilege granted to the Russians, bringing the
competition onto a more equal footing.

In the 1933 agreement, AIOC undertook to sell its products at
a 10 percent discount below the previous prices to the public, and at
a 25 percent discount to the government. These prices were still
competitive with the Russian prices. Under the same agreement,
AIOC undertook to develop the Naft-e-Shah oilfield and build the
Kermanshah refinery near it, so as to supply the bulk of the domestic
requirement of the western and central regions of the country.

The AIOC was unable to supply the domestic requirements of the
country from the Abadan refinery--not because the domestic demand
was so large, but because there were few roads, and transport was
both costly and difficult. Indeed, AIOC was forced to carry the products by rail to Iraq and import them again through Kermanshah, where
they were transported by trucks to the central and northern districts
of Iran. To transport oil to the east and the southeast of Iran, AIOC
had to transport the oil by sea to Karachi in Pakistan and then to
Zahedan, and finally, by different means, to other centers of consumption.[7]

The transportation of oil products was particularly difficult in
Iran, not only because there were few roads but also because of the
very long distance between the centers of production and the centers
of consumption. Oil was produced and refined in the underdeveloped
regions in the south, while the largest demand existed in the more
prosperous central and northern regions. For instance, Tehran, the
largest center of consumption, is 1,022 kilometers (640 miles) from
Abadan. Esfahan, another large center, lies 600 kilometers (370
miles) from Abadan. Other regions in the northwestern and north-
eastern parts of the country are over 1,000 miles from Abadan. Railways were few, inefficient, and slow, and pipelines had been constructed from the production points to export terminals. Until the
mid-1920s the main mode of transporting the cans and drums was by
pack mule or donkey. From the mid-1920s, in the reign of Reza Shah
the Great, the government decided to modernize the transport network and concentrated on making motor transport widespread and efficient. But motor transport required roads, spare parts, roadside
repair facilities, and filling stations. Since none of these existed in
Iran, the government adopted an ambitious road construction policy in
the late 1920s. The existence of roads and motor transport led to the
development of an increasing market for gasoline, which in turn provided the free enterprise with an incentive to build garages and service stations. In short, the availability of supply led to more and
more petroleum-based activities.[8]

MAP 2.1

Transportation of Oil Products, 1928

The first company to use motor transport was AIOC itself. In order to supply its bulk customers in the south, the company employed, before 1928, two trucks each fitted with an 1,800-liter tank. In that year it imported two British-made road tankers each having mounted on it a 3,600-liter tank for service in the south. In that same year "Persaznaft" tried out several road tankers for transporting petroleum products from the Caspian Sea to Tehran. The products carried in these tankers were emptied into tins or other containers at their destination.[9]

In the 1933-36 period the construction of the Kermanshah refinery and the development of the Naft-e-Shah oilfield greatly eased
the demand situation in the western and central regions. By 1937
there were 153 road tankers in operation, of which 105 belonged to
AIOC and the rest to the contractors. A large garage was built in
Kermanshah near the refinery for the servicing and maintenance of
company tankers. Later other garages were built in Mashad and
Shiraz and many other centers for repair work.

Transport by means of road tankers between the centers continued to be the most important method, but in the late 1920s Reza
Shah ordered the construction of a railway line between Ahwaz and
Andimehke, which was later extended to Tehran. The line was completed in 1931. Rail transport at the rate of 0.35 rials per ton-
kilometer was exactly half the cost of transportation by road tankers.[10]
Therefore, wherever railways were built they were used in prefer-
ence to road haulage, and in 1939 the number of rail tankers in service for transporting oil products totaled 150. These tankers were of
German manufacture and each had a capacity of 45,000 liters. Abadan,
the largest refinery in the world, was virtually cut off from the rest
of the country, while Ahwaz, the neighboring city, was connected by
railway to Tehran. There was a need, therefore, to link Abadan to
Ahwaz. In 1939 a 4-inch pipeline with a capacity of 100,000 tons per
year was laid between these two cities. This pipeline is still in use
for transporting aviation fuel from Abadan to Ahwaz.

With the outbreak of World War II the British government considered it essential to open another supply route to Russia. After a
short war, Iran was occupied and Reza Shah decided to abdicate. The
occupation of Iran by the Allied forces brought about an adverse impact on the domestic transport network. The additional road and rail
traffic due to the Allied aid to Russia increased the fuel consumption
considerably, and although the Allied forces used their own transport
for their requirements, a serious strain was imposed on the existing
means. There came about an acute shortage of spare parts and tires,
and freight charges inevitably increased. These factors led to a
shortage in the means of transport needed for Iran's internal requirements. Moreover, the takeover of the trans-Iranian railway system
by the Allied forces imposed a further limitation on the transport of
some of the oil company products, and in certain cases AIOC had to
restrict the sale of petroleum products. Indeed, there is evidence
that oil product distribution in many parts of the country was either
reduced or brought to a standstill.[11]

Ironically, the occupation of Iran by the Allied forces provided
the necessary means for a breakthrough in the oil transport system
of Iran. After the departure of the occupation forces, Iran regained
not only her own road, rail, and pipeline systems, but also the Allied

transport facilities which were either left or sold cheaply to Iran.
In 1946 Iran had 746 railway wagons (an increase of 500 percent in
six years) and 300 road tankers (an increase of 100 percent in six
years). Another important development in the oil transport was the
expansion of the pipeline network. In 1944 another pipeline with a
6-inch diameter and a capacity of 200,000 tons was laid alongside the
existing Tehran-Ahwaz pipeline.[12]

CONSUMPTION AND IMPORTS

An interesting feature of the Iranian oil industry in this period
was that a large part of the requirements of the country was imported
while at the same time   Iran was actually exporting crude and refined
products. The data on the consumption and imports in this period are
hard to find, and what there is must be viewed with caution. For instance, although the most comprehensive set of data available comes
from the Customs and Excise Office in Iran, it is quite possible that
a large volume of oil products were not registered with the Customs
Office. The other main sources of statistics, the Ministry of Finance
and the Foreign Trade Statistics of Iran, differ from those of the
Customs and Excise Office. The Iranian government during this period does not seem to have shown any great interest in compiling
reliable and accurate statistics, and this may well be the answer to
some of the discrepancies. Even British Petroleum is not very sure
with regard to these magnitudes, and refers to its own data as "unofficial." In general, the author is led to believe that the Customs
and Excise Office's data are more reliable than the others, and these
have been used in this selection.

Table 2.2 refers to the 1912-27 period. In this period only
gasoline and kerosene are considered, as the sale of other products
was very small and unrecorded. The sale of kerosene was far more
important than that of gasoline in the whole of this period. However,
after 1923, with the greater improvement of road building and expansion of motor transport, the sale of gasoline rose greatly. In 1924
the ratio of kerosene sales to gasoline was 2:1; by 1927 this ratio
was just under 4:1.

APOC sales in Iran were substantially below the recorded
levels of imports. In 1914 only 12 percent of the total sales were
domestically produced and over 88 percent were imported. In 1921,
82 percent were imported and in 1927 the level of imports was 76
percent of total sales. Table 2.3 shows the source and magnitude of
Iranian imports during this period.

TABLE 2.2

Iranian Consumption of Gasoline and Kerosene in Selected Years
(cubic meters)

Year	Gasoline		Kerosene		APOC	Total	
	APOC Sales	Imports	APOC Sales	Imports		Imports	Sales
1912	--	--	--	44,563	--	44,563	44,563
1915	112	--	3,988	41,536	4,100	41,536	45,636
1918	35	--	2,537	21,006	2,572	21,006	23,578
1921	422	400	4,514	22,524	4,936	22,924	27,860
1924	820	1,032	4,780	33,561	5,600	34,593	40,193
1927	4,682	8,085	9,713	38,296	14,395	46,381	60,776

Source: J. Bharier, Economic Development in Iran, 1900-70 (London: Oxford University Press, 1971), p. 160.

TABLE 2.3

Iranian Import of Oil Products in Selected Years
(metric tons)

Year	Soviet Union	United States	Great Britain	Other[a]	Total[a]	Total (thousands of cubic meters)
1928	50,840	316	2,330	--	53,486	62.0
1930	53,430	215	3,040	--	56,685	65.7
1932	54,840	48	1,620	--	56,508	65.5
1934	49,080	396	580	--	50,056	58.0
1937	40,107	405	3,959	87	44,558	51.7
1938	14,740	505	8,047	100	23,391	27.1
1940	1,777	903	82	187	2,949	3.4
1942	2,262	4,249	178	12	6,701	7.7
1944	17,669	407	32	2	18,110	21.0
1946	53,876	2,094	877	202	57,049	66.2
1948	9,505	343	4,442	394	14,684	17.0
1950	223	1,152	3,608	64	5,047	5.8
1951	--	4,859	935	634	6,428	7.5

[a]Includes Iraq, Hungary, India, Germany, and other countries.

Source: Iranian Customs and Excise Office, Annual Reports, 1927-51.

The outstanding feature of Iran's imports during this period was the importance of Soviet oil. In 1930, of the total imported oil, 94 percent came from Russia. Indeed, throughout most of the period until 1938, Russian oil accounted for over 90 percent of the total imports. Although the import of petroleum products was drastically reduced after 1938, Russian oil still constituted a large portion of the total imports in most years.

Table 2.4 brings together all available data on production, exports, consumption, and imports for the 1912-51 period. It is evident that despite large production and exports, Iran was still obliged to import oil products. In 1922 total consumption was just over 1 percent of total exports; in 1932 and 1942 it was 0.7 and 3 percent respectively. The corresponding ratio was 6.7 percent for 1951.

TABLE 2.4

Production, Exports, Consumption, and Imports
for Selected Years
(metric tons)

Year	Production	Imports	Total Consumption	Exports	Imports as Percent of Consumption
1912	82.0	38.4	38.4	37.0	100
1917	911.8	45.2	48.1	623.0	94
1922	3,006.5	23.4	27.4	2,604.0	72
1927	5,443.7	40.0	52.5	4,754.4	77
1932[a]	6,449.2	56.5	40.0	6,006.3	--
1937	10,329.8	44.5	162.3	9,302.0	28
1942	9,546.7	6.7	265.2	8,030.2	2
1947	20,518.9	34.6	595.9	17,725.1	6
1951	16,885.7	6.4	941.0	14,032.3	0.7

[a]In 1932 imports are larger than consumption implying some re-exports to neighboring countries.

Source: NIOC and Customs and Excise Office of Iran.

Two important points emerge from our analysis. First, imports constituted a large portion of the Iranian consumption of oil products until 1934, after which AIOC's domestic sales gradually expanded, particularly due to the construction of the Kermanshah refinery. Second, Russian oil constituted a large portion of the total import in most years.

Although the rate of import of oil products into Iran may at first sight appear inexplicable, one may in fact find a rather simple explanation. The Iranian transport network was small, inefficient, and costly to operate, and the Iranian market was a small one. Although until 1933 the Iranian prices corresponded to international prices, the high cost of distribution and the small size of the market did not make the sale of domestic products a financially attractive proposition to AIOC.[13] AIOC, being a commercial entity, was not prepared to spend vast sums on the expansion and modernization of the distribution and transport networks in Iran. It was concerned with keeping the Iranian government content by gradually expanding its domestic activities in the country. Indeed, one could say that the

Russians were doing AIOC's job by providing oil products to northern
and central districts of Iran. However, toward the end of the 1930s
the Russian hold on Iran, both in economic and political terms, had
weakened considerably, and AIOC gradually started to comply with
its obligations of the 1933 Agreement by catering for the larger part
of the domestic oil consumption of the country.

SUMMARY AND CONCLUSION

The Iranian oil industry in the period before 1951 was a disintegrated foreign-oriented industry. The domestic supply network
consisted of a few road tankers and railway wagons. The domestic
demand, small as it was, depended on oil imports, particularly from
the Soviet Union, until the late 1930s. AIOC did little to integrate
itself into the national economy. In short, forward linkages through
the supply of products was virtually nonexistent.

NOTES

1. Since the 1951 nationalization the capacity of Abadan refinery has dropped substantially. In 1974 Abadan was producing about
430,000 barrels per day.
2. For details of the operation of the Abadan refinery and other
refineries, see Part III, chaps. 8, 11.
3. Ibid.
4. Ibid.
5. The only information available on this competition is in the
Presidential lecture of E. Ettehadieh at the Iranian Petroleum Institute, "The Haulage of Petroleum Products in Iran," Bulletin of the
Iranian Petroleum Institute 1, Tehran, 1957. In Persian.
6. Unfortunately no details of the Distribution Department's
activities for the period before 1951 are available. Dr. Ferrier, the
British Petroleum Company historian, informs the author that the
records were left in Iran after the Iranian takeover and that the
Iranians appear to have destroyed them.
7. Ettehadieh, op. cit.
8. G. Naamati, A Social and Economic Investigation of the
Effects of Petroleum Product Distribution in Iran (Tehran, 1967),
pp. 164-74. In Persian.
9. Ettehadieh, op. cit.
10. It may be interesting to compare the costs in 1931 with those
in 1973. In 1931 the cost was 0.35 rials for railways and 0.70 rials
per ton-kilometer for road tankers; in 1973 the comparable figures
were 1.12 and 1.25 rials per ton-kilometers. (See Chapter 9.)

11. Naamati, op. cit., p. 210.

12. In 1974 the U.S. government announced that it wishes to negotiate with Iran for the repayment of $36 million worth of equipment, machinery, and transport facilities left in Iran by the U.S. armed forces. The Iranian government in return produced a bill worth $365 million which the U.S. government must pay Iran for its usage of the railway system and other facilities and for the damages. Reported in Kayhan International, Overseas Farsi edition, Jan. 11, 1975.

13. Dr. Ferrier of British Petroleum informed the author that no separate financial statements were prepared by the company on the profitability of domestic activities of AIOC in Iran.

PART

II

IRAN IN A
WORLD OIL
SETTING

3

THE IRANIAN
NATIONALIZATION AND
THE CONSORTIUM
AGREEMENT

In order to understand the background in which the Iranian Consortium came into being, one has to examine two major issues: the nationalization of Iranian oil and the situation of the international oil industry in the early 1950s.

NATIONALIZATION OF THE
IRANIAN OIL INDUSTRY[1]

The causes of conflict between the Iranian government and the Anglo-Iranian Oil Company have been discussed in detail in Chapter 1, but it is generally agreed that these disputes were not the major cause for the nationalization. World War II had increased the suffering and hardship among the poverty-stricken majority, who saw the occupation forces of the Allied Armies perpetuating the political domination of the British government and AIOC over Iran. The need for oil for the war had made all the major powers conscious of the necessity of securing their sources of supply for the future. The interests of the individual oil companies associated with the Allied powers were discussed by their parent countries at the Tehran Conference of 1943. After the withdrawal of American forces, the British left Iran, but the Stalin regime refused to withdraw its army from the northern part of Iran. As the price for withdrawing its troops, Russia extracted a 50-year oil concession from the Iranian Premier Ghavam-e-Saltane, who headed an Iranian delegation to Moscow to negotiate a settlement. An Irano-Russian company was to be formed to explore and exploit the oil of the northern provinces. Russia was to own 51 percent interest during the first 25 years and 50 percent interest during the second 25 years. The profits were to be divided in proportion to

41

ownership.[2] The Russian forces withdrew and Ghavam never submitted the agreement to the Majlis (Parliament) for ratification.
Some believe he never intended to do so. As with many other wars,
the occupation of the country united the people. With the departure
of foreign troops AIOC became once again the focus of opposition.
The fact that AIOC's profits had risen tenfold compared to a fourfold
increase in Iranian royalties in the 1944-50 period added a sense of
urgency to the parliament's demand for a revision in the structure of
the concession. The nationalistic sentiments of the people, stirred
by the Soviet-backed Tudeh Party (the Iranian Communist Party),
forced the government to introduce a brief but significant bill which
became law in October 1947.[3] The law proposed a ban on the award
of any oil concession by the government without prior approval by the
Majlis. The Majlis, therefore, declared the Russian concession null
and void, and instructed the government to investigate and renegotiate
the terms of the AIOC concession, declaring its belief that Iran was
being deprived of its rightful share of the oil revenues.

At the same time, the principle of 50-50 profit sharing was
accepted in Venezuela, and an agreement to that effect was signed in
1947. In 1948 AIOC submitted the so-called "Supplemental Agreement" to the Premier General Razmara. This agreement provided
for an increase in the royalty from 4 to 6 shillings (gold) per ton,
with a tax commutation increase of ninepence to 1 shilling. Under
the agreement the Iranian oil revenues would have increased in the
manner shown in Table 3.1.

TABLE 3.1

Iran's Oil Income Under the Existing and
New Agreements, 1948-50

Year	Iran's Actual Oil Revenues	Oil Revenues under the Supplemental Agreement
1948	£ 9,100,000 (0.37)	£18,600,000 (0.75)
1949	13,400,000 (0.50)	22,800,000 (0.85)
1950	16,000,000 (0.50)	30,000,000 (1.00)

Note: Figures in parentheses represent Iran's revenues in
pounds sterling per ton of oil produced.

Source: L. Nahai and C. L. Kimble, The Petroleum Industry of
Iran (Washington, D.C.: Department of Interior, Bureau of Mines,
1963), p. 17.

In the opinion of the company's chairman, the agreement was the most advantageous granted to any producing country in the Middle East. The Supplemental Agreement would, he argued, give Iran the same benefits as a 50-50 profit-sharing agreement. [4]

The prime minister was satisfied with the agreement and submitted it to the Majlis for ratification. The Majlis ordered the Oil Committee, headed by Dr. M. Mossadegh, to investigate the agreement and report back. The committee's report was unfavorable and the agreement was rejected. In 1949, the Arabian-American Oil Company (ARAMCO) agreed to sign a 50-50 profit-sharing agreement with the Saudi Arabian government. This new development in a neighboring country led to further hostility toward AIOC. AIOC claimed that early in 1951 they offered Razmara a proposal to negotiate for a 50-50 profit-sharing settlement, but the premier wished it to be kept a secret and indeed, he never submitted it to the Parliament. The company chairman had this to say about the offer:

> Despite the company's endeavour to persuade the
> Prime Minister to make known in Iran both the
> company's offer to re-open negotiations for a
> 50/50 profit-sharing settlement, and its action
> to undertake to make advances, General Razmara
> refused to do so and maintained the closest
> secrecy on these matters. [5]

Whether the offer of a 50-50 profit-sharing settlement was actually made to Razmara can never be factually determined. On the one hand, it is clear that Razmara was in favor of a settlement with the company; indeed, he accepted the Supplemental Agreement subject to ratification by the Majlis in 1949, which indicated his wish to avoid a confrontation with the company. On the other hand, it is reasonable to expect that the company was prepared to negotiate such a deal with Iran, as in that period many companies appreciated the benefits of profit-sharing settlements. One can only speculate that if Razmara was offered such a deal his reasons for not submitting the proposals to the Majlis were that he knew that the proposal was not satisfactory enough to be ratified by it, and he dared not risk another rejection of his proposals by Parliament. On February 19, 1951, Dr. Mossadegh submitted a draft project for the nationalization of the oil industry and all the AIOC properties in Iran. Razmara declared his opposition to the nationalization project by saying that Iran could not, for technical, economic, and political reasons, nationalize her oil industry. Four days later he was assassinated. Dr. Mossadegh assumed power on April 28, 1951, and on April 30 the Majlis approved the Mossadegh proposal to nationalize AIOC's properties. And thus

the successful boycott of the Iranian oil by the forces of Western
interest started.

Mossadegh was an emotional man with little foresight. Though
initially popular, he was a ruthless dictator who would quash all opposition with little mercy. After the International Court of Justice in
The Hague had rejected the British government's request to intervene
by declaring that it had no jurisdiction over Iran, Mossadegh asked
for large dictatorial powers from the Majlis. The latter gave him a
unanimous vote of confidence, but the Senate hesitated. Mossadegh
persuaded the Majlis to vote the Senate out of office. He further dissolved the Supreme Court and asked for a new electoral law. His
wish was granted by the Majlis, but when the Majlis hesitated to ex-
tend his dictatorial powers to rule by decree for another year, his
emotional speeches incited large and violent demonstrations against
the Majlis, until he got his way. When he challenged the authority of
the Shah the majority of the Majlis Deputies turned against him, so
in August 1953 he ordered a plebiscite to dissolve the Majlis. Meanwhile the Shah left the country and named General Zahedi prime minister. Mossadegh did not recognize the Shah's authority to dismiss
him, and continued in office. Although Mossadegh was still in power,
the effect of his promises was wearing off. He had no success in
selling Iran's oil to America or Russia, or to other European countries. Economic crisis, combined with a sense of purposeless struggle from within, together with the external forces of Western interest,
brought about Mossadegh's downfall. General Zahedi ousted Mossadegh in a coup d'état in September 1953. The Shah returned, and with
the help of American mediation, an agreement was reached with a
consortium of international oil companies in 1954.

INTERNATIONAL OIL INDUSTRY AFTER
WORLD WAR II[6]

The late 1940s and early 1950s proved to be a turning point for
the international major oil companies.[7] Although these companies
were still controlling a large portion of the international oil industry,[8]
there were some unexpected erosions in their dominant position.

After the "As Is" agreement of 1928, the majors kept in close
and regular contact with each other, arranging their international
sales and production policies in such a way as to maintain prices at
agreed levels. But toward the end of the 1940s and the early 1950s,
there were some new developments in the international oil industry
which shaped the oil industry into a different form in the next two
decades. These developments can be summarized as follows:

(a) The Venezuelan and Saudi Arabian Agreements of 1947 and 1949, with their 50-50 profit sharing, were a most important step in changing the company-government relationship. Although these agreements might have been expected to have had an adverse effect on the companies' profits, they did in fact turn out to be less costly for the companies. This was because the governments of the United States and the United Kingdom agreed (in 1951 and 1954 respectively) to allow the payments to the producing countries to be credited against the companies' domestic tax liabilities. Indeed, the oil industry went full circle insofar as the payments to the producing countries were concerned. The payment system began on a profit-sharing basis until the early 1930s; thereafter the principle of tonnage royalty was introduced, although some profit-sharing provisions were allowed for. The 50-50 profit-sharing agreements signifies a return to the D'Arcy principle of oil revenues being based on company profits, but there were great differences in these two types of profit-sharing arrangements, the most important being the increase from 16 to 50 percent of profits as payments to host countries.

(b) Although the Iranian nationalization was not the first of its kind (Mexican Eagle was nationalized before World War II), it was certainly the most important. Iran was the largest Middle East oil producer and AIOC was a major international oil company, so that nationalization of AIOC's properties created a turmoil in the international oil industry. The companies began to realize that their relationship with the host governments was undergoing an unprecedented change. Indeed, the Iranian nationalization closed the ranks of the companies and the fear of further nationalization was, perhaps, the most important motive for solidarity of the U.S. companies with AIOC in refusing to handle the nationalized Iranian oil.

(c) After the Indian price war of 1927, three largest international oil companies, AIOC, Shell, and Jersey, met at Achnacarry and signed the 1928 "As Is" agreement. The agreement was to maintain the status quo and to avoid price wars. The prices were based on a "single basing system"; because the United States was the largest exporter of oil, the prices were based on the f.o.b. prices of the Gulf of Mexico, plus the agreed freight cost to the port of delivery. The so-called 'Gulf plus" pricing system had some great advantages for the majors. First, there was a common yardstick to which all the companies could adhere and thus avoid competition in various markets. Second, the cost of crude was irrelevant to the prevailing prices. The latter system was to ensure that no company could sell at lower prices when it discovered a low-cost source of crude. Thus the fact that the Persian crude costs were much lower than those in the United States meant that AIOC was making particularly large

profits and would not be able to lower its prices in competition with
the other oil companies. World War II created some difficulties for
the companies. The British Navy stationed in the Indian Ocean was
unhappy about paying the Gulf of Mexico prices, plus the freight
charges, while it received its requirements from Iran. Under pressure from the consuming governments, the oil companies agreed to
adopt a "dual-base pricing system," with the Persian Gulf as the
second basing point. The companies, however, insisted that the
Persian Gulf f.o.b. should equal those of the Gulf of Mexico; thus the
only reduction in price was the lower transport costs for recipients
nearer to the Persian Gulf.

The relative decline in the Middle East prices after World War
II has been the subject of picturesque theories which we do not need
to go into. [9] One plausible explanation of this downward trend can be
provided through the study of the change in the institutional arrangements of the major oil companies which had worked so well in maintaining stable prices during the 1930s and 1940s.

Before the war, the vertically integrated nature of the majors
and their dominance of the international oil industry had made it unnecessary for a crude market to be developed. Since the refineries
were located near the centers of production, only refined products
were traded. After the war, the main consuming countries brought
pressure on the companies to construct refineries close to the consumption centers. This would have saved them foreign exchange
while at the same time providing them with a certain sense of security of supply. Between 1947 and 1951 the refinery capacity in
Western Europe tripled; thus the amount of crude entering international trade substantially increased and the relative importance of
trade in products declined. For the refining activities to operate
efficiently, the management had to be given incentives to show profits,
and since for a group of companies within an integrated network the
prices at individual stages of operation were transfer prices, not
influencing the total integrated profits (before tax), it was desirable
to offer oil at lower prices to the refining affiliates. At this point in
time the United States had become a net importer of oil and the Middle East was supplying the bulk of the Eastern Hemisphere's requirements. Thus from the point of view of the major consumers a decline in the Middle East prices was most advantageous.

The extent to which competition was responsible for the decline
in Middle East prices is a highly controversial subject. While it may
not be possible to explain the level and structure of prices in terms
of a competitive market ruled by the free play of supply and demand,
it is equally incorrect to say there was no competition at all, slow
and imperfect though it might have been. The institutional arrange-

ments, which had enabled the oil companies to control the prices
over the past two decades, had undergone a significant structural
change. As the 1950s wore on, the competition in refined products
became keener and new firms (private and state-owned) appeared on
the scene, creating an unstable arm's-length crude market. The
major oil companies had no interest in causing the collapse of international prices, but as their hold on the market weakened they attempted to increase their market share of product sales individually
through competition or discounts.

At the time (and even today) the Russians and independent (non-
integrated) oil companies were blamed for the weakening of the price
structure. The Russians were said to have offered oil at cheaper
prices as a part of their "cold war" strategy, while independents were
alleged to be economically irrational for trading in cheap crude.
There is, however, little evidence to prove that the Russians and
independents were the major forces behind this downward trend. To
be sure, they had some impact on the prices, but the collapse of the
Middle East prices is only explicable in terms of the changes in the
organizational structure of the major oil companies.

The end result had far-reaching consequences for the host
countries. The posted prices had then become tax reference prices
under the 50-50 profit-sharing formula, used for calculating the payments to the producing governments. The posted price of Persian
Gulf crude which was $2.22 in 1948 had fallen to $2.04 per barrel by
1957. In February 1959 the prices dropped to $1.86 per barrel, and
in August 1960 it was further reduced to $1.78 per barrel. The in-
ability of the oil companies to stop erosion of the posted prices was
the cause of the creation of the Organization of the Petroleum Exporting Countries (OPEC) in 1960 (see Chapter 5).

Participants of the Iranian Oil Consortium

After three years of political and economic turmoil in Iran, the
way became clear for an agreement with a consortium of interna-
tional majors. The Iricon Group is a consortium of independent
American oil companies whose members are American International
Oil Company, Atlantic Richfield Company, Charter Oil Company,
Continental Oil Company, and the Standard Oil Company of Ohio.
With regard to the Iricon Group, P. H. Frankel, an oil consultant,
had this to say:

> For the additional members of the Iranian Consor-
> tium, their minor share proved to be an exceedingly
> profitable investment and the prospectus which was

drawn up by a large firm of Chartered Accountants
showed clearly, for all concerns, that even after
the compensation to be payable to AIOC, any stake
in that venture was a license to print money. [10]

Needless to say, the U.S. Justice Department gave an assurance to
the five U.S. majors that their joint participation in the Iranian Consortium would not constitute an illegal restraint of trade.

Company	Percentage Share	Nationality
British Petroleum	40	British
Royal Dutch/Shell	14	Anglo-Dutch
Standard Oil of New Jersey	7	United States
Standard Oil of California	7	United States
Texaco	7	United States
Mobil	7	United States
Gulf	7	United States
CFP	6	French
Iricon Group	5	United States
Total	100	

Interrelationship Between the
Consortium Participants

One of the original Iranian objections against the Anglo-Iranian
Oil Company was that Iran was not satisfied with leaving her oil in
the hands of a single company. The placing of Iranian oil into the
hands of several participants--16 in all--may seem on the surface
to have solved the Iranian objection. But a closer look at the interrelationship between the Consortium participants reveals that the
real change in the overall control of Iranian oil by foreign oil companies was negligible. The seven majors, plus CFP, controlled
95 percent of the Iranian oil. Table 3.2 will help to show that so
far as Iran was concerned, none of the participants could act independently from the others without jeopardizing its own interests
elsewhere.

Table 3.2 shows the extent of the interrelationship between the
major concession holders of the Middle East. It gives the impres-
sion that a monopoly was replaced by an international cartel of the
major oil companies in Iran. Since the nature of any arrangement
between these companies would be kept as secret as possible, much

TABLE 3.2

Interrelationship Between the Major Oil Companies in 1966
(percentage ownership of the Middle Eastern operations)

	Abu Dhabi Marine Areas	Kuwait Oil Company	Iranian Consortium	Iraq Petroleum Company	Arabian-American Oil Company
B.P.	66-2/3	50	40	23.75	--
Mobil	--	--	7	11.875	10
Socal.	--	--	7	--	30
Texaco	--	--	7	--	30
Jersey*	--	--	7	11.875	30
Gulf	--	50	7	--	--
Shell	--	--	14	23.75	--
CFP	33-1/3	--	6	23.75	--
Total	100	100	95	95	100

*Also known as Standard Oil of New Jersey, Esso, Humble Oil, and more recently Exxon.

Source: E. T. Penrose, The Large International Firm in Developing Countries—The International Petroleum Industry (London: George Allen and Unwin, 1968), p. 151.

of the evidence for their existence would have to rest on inference from the observed behavior of the companies. The Federal Trade Commission in its report, "The International Petroleum Cartel," made a comprehensive attempt to collect the available evidence and, as the title of their report implies, concluded that the designation "cartel" was appropriate. It is outside the scope of this chapter to argue whether or not the existence of a cartel was necessary for the maintenance of price stability and the prosperity of the international oil industry, but it seems incontestable that the rate of supply was subject to some sort of jointly controlled planning. Planning of supply was only possible because of the two inherent characteristics of the major oil companies--the joint control/ownership of crude oil resources and the vertical integration within the oil industry. In this way each international oil company was able to adjust its output of crude oil to its planned sales of crude and refined products with a high degree of effectiveness. Professor Penrose writes:

> Moreover, it is not at all clear that price movements played any role at all in the supply adjustments consequent on the virtual cessation of

production in Iran during the Abadan crisis from
1952-55 and in subsequent re-absorbtion of the
Iranian output after the settlement.[11]

Clearly, without a coordinated system of production the disappearance
of the crude exports from the largest Middle East oil producer would
have had a great impact on international petroleum prices.

Although one may argue that the Consortium Agreement replaced
one major oil company by eight majors, and that Iran may have been
in a weaker position compared to the past, it is fair to say that the
broader nature of the participants helped to expand the Iranian oil industry at a faster rate.

THE 1954 CONSORTIUM AGREEMENT

Having won the struggle, AIOC and the British government
could not be expected to grant more than they had previously offered.
They agreed to recognize the principle of nationalization insofar as
it did not affect their normal rights and privileges. In the words of
Professor Stocking, "It (the Consortium Agreement) gave to the
Iranians the shadow of what they sought, while retaining for the British the substance of what they had."[12] The Consortium Agreement
contains provisions recognizing that the entire assets belong to Iran,
but also clearly lays down the right of the foreign oil companies to
operate there. Indeed, it is a fallacy to argue that the Consortium
Agreement was different from the agreements signed by the other
majors with other oil-producing countries of the world. NIOC and
the Iranian government refer to the Consortium Agreement as an
"agency contract" in their official publications, because they claim
that the oil reserves and the assets belong to Iran and the Consortium
is only operating these assets for Iran. But the Consortium Agreement is in substance another concession of the 50-50 profit-sharing
type, signed by other Middle East producers. The only difference
between the other 50-50 type agreements and the Consortium Agreement is concerning the ownership of the reserves and assets. In the
former case, the reserves and assets belonged to the concessionaire,
which would return them to the host country after the concession had
ended, while in the latter case the oil reserves and assets belonged
to Iran, but the concessionaire was allowed to operate them until the
end of the concession period.[13] In no case could the host country
interfere in the running of the concession. Thus, basically there
were no differences in the two kinds of concessions.

Details of the 1954 Agreement

The area covered by the new agreement was about the same
as the one in AIOC's concession. It included 100,000 square miles,
extending from a point about the middle of Iran's western borders in
a southeasterly direction to the north of the Persian Gulf, an area
approximately 750 miles long and varying from 50 to 200 miles in
width. It embraced all the fields AIOC had operated except the
Kermanshah oilfield, which was taken over together with the Kermanshah refinery by NIOC. The signatories to the agreement were
NIOC, representing the government of Iran, and the Consortium.
NIOC was given the ownership of all producing, refining, and auxil-
iary installations operated by AIOC and all the future facilities of the
Consortium members.

The agreement called for the creation of two operating companies, Iranian Oil Exploration and Producing Company and Iranian
Oil Refining Company. These companies were to be registered in
Iran but established under the laws of the Netherlands. The function
of these companies was to operate the Iranian oil industry on behalf
of NIOC. Each company has seven directors, two of which were
nominated by NIOC. A token right of inspection and audit was also
awarded to NIOC under Article 4 of the agreement.

The operating companies are not profit-making entities and
thus do not directly engage in selling or export of oil. This function
was undertaken by the trading companies set up by individual consortium participants, which bought the oil at wellhead from NIOC.
The Iranian Oil Exploration and Producing Company delivers the
oil for the account of the trading companies to export terminals or
to the Abadan refinery. The Iranian Oil Refining Company refines the
requirement of the trading companies at Abadan for a fee of 1 shilling
per cubic meter. The crude or product is usually sold, at the loading port, by trading companies to one of their affiliates at transfer
prices (intracompany accounting prices).

The Consortium members have also formed two companies in
London--one, Iranian Oil Participants Limited, to hold the shares
in the two operating companies and the other, Iranian Oil Services
Limited, for the provision of supply and personnel for their operations. Under the new agreement, NIOC was to take over the "nonbasic operations" which included health, housing, and education.[14]
The agreement was to run for 25 years with the provision of three
five-year renewals at the Consortium's option.

Profit Division Method

Payments to the Iranian government take the form of royalty payments and 50-50 net profit division. Royalty was fixed at 12.5 percent
of total production. Only Iran and Iraq obtained provisions from the
majors that this 12.5 percent could be payable either in kind or in
cash at posted prices. Other Middle East oil-producing nations were
bound to sell this 12.5 percent for cash. The 50-50 profit-sharing device as operated in Iran and other producing nations, means that the
net profits of the producing stage should be divided equally between
the Consortium and NIOC.

The 50-50 profit-sharing device meant that the oil-producing
country's per barrel receipts nearly tripled overnight. Although it is
not certain whether the idea originally initiated from the host countries
or from the oil companies themselves, it is nevertheless a fact that
the oil companies encouraged the idea. They did so for two reasons:
first, the increased payments to the host countries was almost entirely offset by tax credits in their parent countries, and second, it
created a good deal of goodwill in the host countries, temporary
though it may have been. In fact this generosity cost them almost
nothing. Table 3.3 shows the mechanics of the tax credit system as
well as the principles underlying the 50-50 agreements.

Table 3.3 shows that the companies did not contribute more than
7 cents per barrel to the increase in host countries' receipts. At the
same time the royalty payments were completely avoided, since the
royalty was treated as part of the 50 percent income tax liability. The
latter issue became the cause of the first major confrontation between
OPEC and the companies in the mid-1960s, as we shall see in Chapter 5.

A crucial point which must be emphasized at this stage is that the
profit-sharing principle had in fact nothing to do with the oil companies'
profits. The oil company profits are based on the net integrated income of a group of affiliates involved in production, transportation,
refining, and marketing of oil products. The 50-50 arrangement was
concerned only with an equal sharing of profits attributed to the production stage.

AIOC's Terms of Compensation

Under the Consortium Agreement Iran was to pay the British
Petroleum Company (formerly AIOC) the sum of 25 million pounds
sterling as compensation for nationalization of its assets. The figure
represented the difference between what the company claimed were
its losses from July 1951 until the 1954 Agreement, and the additional
sums due to Iran under the 1949 agreement. Iran was to pay this £25
million in ten yearly installments.

TABLE 3.3

Fiscal Impact of the 50-50 Agreements

	Host Country Calculations		U.S. Calculations	
	Before	After	Before	After
Price per barrel	$2.00	$2.00	$2.00	$2.00
Production cost[a]	0.60	0.60	0.60	0.60
Royalty	0.25	0[b]	0.25	0
Depletion allowance	n.a.	n.a.	0.394[c]	0.45[c]
Gross taxable profit	1.15	1.40	0.756	0.95
Income tax at				
50 percent	n.a.	0.70	0.378	0.475
Foreign tax credit	n.a.	n.a.	0	0.475
				$(+ 0.225)$[d]
Government take	0.25	0.70	0.378	0

	Before	After
Total payments to governments,		
per barrel	$0.628	$0.700
United States	0.378	0
Host	0.250	0.700
Cost of the new agreement to the		
oil companies (per barrel)	n.a.	0.072

n.a. = not applicable.

[a]Including the opportunity cost of capital.

[b]Royalty was treated as advance payments toward income tax and not deducted as cost of operations.

[c]Equals 0.225 times the price minus the royalty.

[d]"Excess" credit arises because foreign tax exceeds U.S. tax; excess can be used to offset other tax liabilities on foreign source income only under restricted conditions.

Source: Based on the lecture notes of Dr. T.R. Stauffer at Harvard University.

In return for the 60 percent British Petroleum gave up to the other companies, it was to receive an immediate sum of £32.4 million, plus 10 cents per barrel on all exports until a sum equivalent to £510 million had been reached. The terms were duly carried out during the ensuing years. It was stated that AIOC had been fully indemnified by the Iranian government and the other members of the Consortium, in accordance with the claims it originally submitted, and had emerged from the nationalization crisis in Iran with reasonably

little loss. It was also stated that AIOC's producing subsidiaries in
Iran received about £16 for each £1 share for the surrender of their
rights in Iran.[15]

Collaboration Among the Consortium Participants

It was shown in the previous pages how interrelated the major
oil companies are with regard to crude liftings, but their method of
liftings from Iran was a tightly held secret that was blown open by the
U.S. Senate foreign relations subcommittee on multinational corporations in early 1974.[16]

Nominations by the companies set what was known as the Consortium's "aggregate programmed quantity" or APQ. This was arrived
at through a rigid formula agreed to by the Consortium participants
in 1954. To arrive at APQ, nominations by each company were trans-
lated into a production total required to provide each with its desired
volume based on its equity share. All nominations were then listed in
a descending order of magnitude until a cumulative 70 percent equity
level was reached. At that point the so-called APQ volume for the
year was decided. Table 3.4 gives an example for the year 1966.

TABLE 3.4

Method of Determination of Production Levels

Participant	Equity Share, Percent	Cumulative Shares, Percent	Total Program (thousands of b/d)
Iricon	5	5	2,030
British Petroleum	40	45	2,027
Shell	14	59	2,027
Mobil	7	66	1,964
CFP	6	72	1,945
Esso	7	79	1,890
Texaco	7	86	1,712
Gulf	7	93	1,700
Socal	7	100	1,644

Note: In this particular instance, CFP's nomination brought the
cumulative total to 70 percent of the equity, so its nomination of
1,945,000 barrels set the Consortium's program for that year.

Source: Petroleum Intelligence Weekly, April 29, 1974.

As one might expect, the crude short companies such as Mobil, Shell, CFP, and Iricon, consistently voted for the highest production levels. Conversely, the crude long firms such as Socal, Exxon, Texaco, or Gulf with large production in Saudi Arabia and Kuwait, consistently voted lower. The low nominators had simply no effect on the decided volume of lifting. British Petroleum was the most important force in determining the volume of liftings; together with any group of small equity holders (adding up to 30 percent), it could impose its will on the others. Table 3.5 provides detailed company-by-company nominations for Iran from 1957-73. The nomination showing the appropriate APQ in each case is underlined.

Table 3.5 brings to light some unexpected peculiarities indicating very close collaboration between the participants. First, nominations had little correlation to the percentage ownership--indeed, BP and Shell, the largest owners, did not nominate particularly higher volumes in comparison with the smaller owners. Second, in many instances one can find exactly equal nominations by participants with different percentage ownerships. For instance, in 1969 British Petroleum, Shell, Mobil, and CFP, with a total ownership of 67 percent of the Consortium, all nominated 3,014 thousands of barrels daily--and in many years this pattern was repeated, which cannot be accepted as coincidence. Finally, the voting pattern was abruptly changed when world crude availability began shifting from surplus to tight supply in 1972 and 1973. After the Kuwaiti cutback in 1972, Gulf became the highest nominator in Iran in 1973, and Exxon, previously a low nominator, raised its bid sharply for both 1972 and 1973.

1973 SALES AND PURCHASE AGREEMENT

The Consortium Agreement was originally drawn to run until 1979. From its inclusion in 1954 to 1971, the agreement maintained its basic characteristics, despite some minor modifications.

The year 1971 was the year of change in the state of the international oil industry. On February 15, 1971, the Tehran Agreement was signed, raising the posted prices of crude oil by the six Persian Gulf members of OPEC. Later in the year other producers followed suit. The Tehran Agreement signaled a major change in the relationship between the consumers and the producers; the buyers' market was transformed into a sellers' market, and the solidarity of the producers began to pay off.

In July 1971 some OPEC members led by Sheik Ahmad Yamani, the Saudi Arabian Oil Minister, demanded equity participation in the oil companies operating in their countries. On December 20, 1972,

TABLE 3.5

APQ Tablings in Iranian Consortium, 1957-73
(thousands of barrels daily)

1957		1958		1959		1960		1961		1962	
BP	750	CFP	827	Iricon	912	Mobil	1,111	Mobil	1,213	Iricon	1,370
CFP	750	Iricon	826	Mobil	890	Iricon	1,023	Iricon	1,201	BP	1,370
Shell	714	Shell	822	BP	850	BP	984	BP	1,192	Shell	1,370
Iricon	700	Mobil	814	Shell	850	Shell	960	Shell	1,192	CFP	1,370
Exxon	690	BP	787	Socal	850	CFP	911	CFP	1,192	Mobil	1,342
Socal	603	Exxon	759	Exxon	841	Exxon	900	Exxon	980	Texaco	1,195
Gulf	574	Socal	724	CFP	822	Gulf	880	Texaco	980	Socal	1,170
Texaco	528	Texaco	700	Texaco	805	Texaco	875	Socal	973	Exxon	1,115
Mobil	528	Gulf	655	Gulf	680	Socal	874	Gulf	940	Gulf	1,000

1963		1964		1965		1966		1967		1968	
BP	1,575	Iricon	1,740	CFP	1,863	Iricon	2,030	CFP	2,274	Iricon	2,698
CFP	1,575	Mobil	1,721	Iricon	1,860	BP	2,027	Shell	2,233	Shell	2,678
Iricon	1,535	Shell	1,708	BP	1,836	Shell	2,027	BP	2,219	Gulf	2,571
Shell	1,534	BP	1,680	Shell	1,836	Mobil	1,964	Mobil	2,178	BP	2,568
Mobil	1,521	CFP	1,680	Mobil	1,836	CFP	1,945	Iricon	2,178	CFP	2,568
Socal	1,360	Texaco	1,470	Gulf	1,685	Exxon	1,890	Exxon	2,100	Mobil	2,568
Texaco	1,299	Exxon	1,400	Texaco	1,589	Texaco	1,712	Texaco	2,005	Socal	2,391
Exxon	1,200	Gulf	1,314	Exxon	1,575	Gulf	1,700	Socal	1,973	Exxon	2,363
Gulf	1,068	Socal	1,214	Socal	1,370	Socal	1,644	Gulf	1,871	Texaco	2,363

1969		1970		1971		1972		1973	
Iricon	3,062	Iricon	3,547	Iricon	4,054	Iricon	4,678	Gulf	5,428
BP	3,014	Shell	3,346	Mobil	3,986	CFP	4,645	Iricon	5,342
Shell	3,014	Mobil	3,342	BP	3,973	Exxon	4,638	CFP	5,297
Mobil	3,014	BP	3,329	Shell	3,973	Shell	4,590	Exxon	5,250
CFP	3,014	CFP	3,329	CFP	3,973	Mobil	4,516	BP	5,137
Gulf	2,900	Exxon	3,205	Exxon	3,973	Socal	4,500	Mobil	5,068
Exxon	2,849	Gulf	3,200	Socal	3,767	BP	4,372	Shell	5,055
Socal	2,795	Socal	3,164	Texaco	3,562	Gulf	4,257	Texaco	5,043
Texaco	2,679	Texaco	3,134	Gulf	3,286	Texaco	4,200	Socal	4,857

Source: Petroleum Intelligence Weekly, April 29, 1974, p. 3.

the "model" participation agreement was signed by Saudi Arabia and
Abu Dhabi, providing for an immediate 25 percent participation, and
51 percent control by January 1, 1982.[17]

Iran was not a party to the participation agreement, opposing
it from the outset. This opposition was based on the belief that the
program was unworkable. But the success of the participation program left Iran in a difficult position. The buy-back arrangement of
the participation crude meant that the signatories to the agreement
would have a larger disposable per barrel revenue than Iran, with a
greater political control over the companies. Moreover, the producers had in their possession larger quantities of crude, which they
themselves could export if they so wished. The Shah and NIOC had
to do something to provide Iran with at least as good a deal as the
other producers had received. This opened the talks on a complete
takeover of the Iranian oil industry by NIOC.[18] The negotiations
with the Consortium started in July 1972 and an agreement was signed
in July 1973.[19]

Major Features of the New Agreement

Under the new agreement NIOC will take over the entire operation of the Iranian oil industry. This was the goal of the 1951 nationalization attempt, which was realized only after two decades. This
takeover is by no means a nationalization insofar as expropriation
of the foreign operator's assets is concerned. Although no terms of
compensation in cash terms were discussed, NIOC is obliged to sell
crude to the Consortium members under a 20-year supply agreement.

The sale of crude will be governed by the OPEC-posted prices,
and in this sense the percentage profit share of the foreign operator,
after the deduction of various costs and royalties, is essentially
preserved. The main change which took place was that Iran agreed
to do the Consortium's job by exploiting the crude and transporting
it to the export terminals for delivery to the Consortium tankers.
Moreover, NIOC's production policy has become independent of the
majors and it will in the future produce and export oil in accordance
with national needs and not in conjunction with the international supply
and demand conditions (as the majors had done for years). The
Abadan refinery is taken over by NIOC, but the foreign operator
reserves the right to its previous 300,000 b/d after the products
for internal demand are lifted by NIOC.

An important part of the agreement is Article 6, which describes the new financial arrangements on prices which will ensure
that NIOC will not receive less revenue in terms of dollars per barrel than its Arab neighbors involved in participation programs.

There are four elements included in the prices which the Consortium members will pay for crude delivered by NIOC, either on board ship or for processing at Abadan.   These are

(a) An operating cost of 10-11 cents per barrel.   This may rise when NIOC undertakes secondary recovery methods.
(b) A fully expensed royalty of 12.5 percent of posted prices.
(c) The balancing margin principle, which takes into account the benefits Iran would have received under participation, but allowing for the Consortium companies' depreciation charges on the remaining unamortized net book value of their assets in Iran.   The margin is estimated at 6.5 cents per barrel until the end of 1975, but subject to revision thereafter.
(d) Interest on 60 percent of the capital expenditure provided by NIOC to increase the production capacity of the Iranian oilfields.   This will last only for five years, which embraces the period of major capacity expansion.   The interest payable is estimated to be 0.418 cents per barrel in 1973, but this will increase in future years.   The remaining 40 percent of capital requirements necessary to expand production will be advanced by the Consortium companies as payment for crude oil purchases.

It is estimated that for the nine-month period from March 21, 1973 to December 31, 1973 (the agreement was retroactive to March 21, 1973), the Consortium would pay 55 cents per barrel for the crude purchased from NIOC.   In addition they will pay income tax to Iran on the difference between that figure and the posted prices, which will enable them to retain a tax credit against home-based income tax.[20]

Another important aspect of the new agreement is the production policy of NIOC.   NIOC is to produce 42.5 billion barrels of oil in the 1973-93 period.   Of this figure, 29.3 billion barrels are to be sold to the Consortium members, 6 billion barrels left for domestic consumption, and 7.2 billion barrels for direct exports.   Production would rise from 4.5 million b/d in 1971 to 7.6 (or even 8.0) million b/d per day in 1977.   Together with production from other joint ventures, Iran's installed capacity is expected to reach 8.5 to 9.0 million b/d by then.   Thereafter the production will remain steady for seven years, or until 1984.   From then onward production would fall so that it reaches 1.5 million b/d (one-third of production in 1971) in 1993.   The internal consumption was expected to be around 1.5 million b/d in 1973, implying that the 1993 production level would be in line with domestic consumption.[21]

SUMMARY AND CONCLUSION

The Supplemental Agreement of 1949 offered to Iran the terms
of a contract which would, in the opinion of some experts, yield the
same income to the Iranian government as a 50-50 profit-sharing
agreement. [22]

Dr. Mossadegh, as head of the Parliamentary Commission on
oil, recommended the rejection of this agreement by Parliament. It
was also stated by AIOC that a 50-50 offer was made to Iran, which
never became public. Mossadegh's insistence that Iran should share
50 percent of the net integrated profits of AIOC and not 50 percent of
the net profits at the production stage was a demand doomed to failure. It brought about three years of economic and political crisis for
Iran, ending with an agreement no more favorable than those already
offered to Iran, or those accepted by other Middle Eastern countries.
This emotional and hurried decision entailed for Iran the loss of three
years of oil income, loss of Iranian lead in production, expansion of
production in other countries which were potential competitors, and
a sense of political demoralization and defeat for the Iranian people.
One of the principal architects of the Consortium Agreement, Dr. Ali
Amini, then Minister of Finance, told the Majlis:

> The solution that I bring to you is perhaps not the
> ideal solution . . . but we do not yet have the means
> to compete in the international markets because we
> do not possess a marketing organization . . . if
> anyone is capable of doing better, let him take our
> place. [23]

There is little doubt that if Iran had not nationalized her oil she
would have received exactly the same type of concession agreement
from AIOC as indeed all the other producing nations did. It is important to emphasize that the 50-50 profit-sharing agreements were
not disliked by the majors; first, they would receive tax credits
against their home country taxes, and second, these agreements
were not quite 50-50 as we have seen--the companies evaded royalty
payments by counting the 12.5 percent royalty as part of the income
tax.

The Consortium Agreement did not bring any extra benefits for
Iran, at least in the short run. It brought Iran essentially the same
type of contract as Iran would have obtained anyway. Insofar as the
Iranian objections against a single monopoly (AIOC) was concerned,
the monopolistic structure did not really change. Indeed, Iran was
in a weaker position facing eight majors rather than facing a single
major oil company. As to the nationalization principle accepted by

the Consortium, this did not change the picture at all. The assets
belonged to NIOC, but were exploited as any other concession by the
Consortium. As far as the provisions of the agreement with regard
to NIOC's control of the Consortium operations were concerned (such
as having NIOC representatives on the board of directors of the producing and refining companies and NIOC's right of inspection of
records), Iran was unable to exercise any effective control. Important decisions on the level of production, the expansion of domestic
reserves, and purchase prices were left to the Consortium members.

Although the Iranian nationalization brought no immediate extra
benefits for Iran, it proved to be of great help to the Iranian oil industry in the long run. Whether this would make up for the material
losses, the political turmoil, and the individual suffering that accompanied the nationalization may never be established. A lesson
Iran learned was how dependent her economy was on oil; the mistake
was never repeated again. Perhaps the most significant consequence
of the nationalization was the creation of NIOC. NIOC was the first
national oil company in a major oil-producing country. Its immediate
task of taking over the domestic distribution of the oil products in
Iran, as we shall see in the future chapters, contributed greatly to
the material well-being of the country and its economic development
by providing cheap energy and expanding its distribution network.
As a young company, it observed the operations of the Consortium
and gradually obtained a great deal of experience and know-how. The
birth of NIOC gave rise to the creation of joint-venture and agency
contracts, which, as we shall see in Chapter 4, changed the structure of the relationships between the oil companies and the host
governments.

The Iranian takeover of the Consortium's operations was a
victory for Iran, brought about once again by the personal intervention of the Shah. For the time being, nothing substantial has changed
in the structure of the operations, except perhaps in decision-making
process on production levels. NIOC has contracted a private company called Oil Services Company of Iran (OSCO) which runs the
Consortium's agreement area. The majority of the Consortium's
employees, both foreign and Iranian, are working in the new company which occupies the former Consortium's quarters. At the same
time the Consortium participants are buying from Iran for 20 years
the crude that they would have lifted themselves had the Consortium
agreement not been canceled. The financial terms are also similar
in basis to the old agreement, except for the balancing margin principle.

But the transition periods are always difficult. NIOC will by
the end of the decade run OSCO or integrate it within itself. Furthermore, NIOC will endeavor to invest heavily in a search for oil to

increase its recoverable reserves. Methods of secondary recovery and possibly tertiary recovery will be employed by NIOC to make a fuller utilization of the existing wells. In 1973 the crude recovery rate in southern oilfields was 15-35 percent depending upon the oil-field. NIOC planned to spend $500 million in the next three years to increase the recovery rate. This ambitious plan points to the problems of NIOC's role as a national oil company in comparison to multinational firms. Secondary recovery through gas or water re-injection is a costly venture which is not ordinarily undertaken by the major oil companies on a large scale.

Most important of all, the new agreement was the first step toward the exclusion of the major oil companies from production operations and making NIOC a major international oil company. Insofar as the 20-year supply contract is concerned it is worth noting that the events in the past few years in the oil business have shown that the agreements of this kind are not too important when the questions of national sovereignty are considered. Iran will no doubt revise or cancel the agreement if she finds it beneficial to do so. But with the Consortium operating the southern oilfields, a total cancellation would not have been possible without a great deal of international friction and a possible shutdown of the installations. Thus the 1973 agreement must also be viewed as a strategic act to provide a smoother exclusion of the major oil companies.

With regard to the financial terms, the new agreement guaranteed Iran per barrel income comparable to the neighboring countries, through the balancing margin principle. This arrangement meant that Iran did not have to engage itself in negotiation for an increase in her income. Iran could simply sit on the fringe and allow the other producers to increase their take (through participation or otherwise) and simply reap the benefits.

It was mentioned earlier that the new agreement has empowered NIOC to decide on production levels. This might on the surface seem to have provided Iran with a chance to substantially increase its income if the need arises. This, however, is not the case; the Iranian production has nearly reached its peak and it is unlikely that the present production level of around 5.8-6.0 million barrels per day can be improved by more than an extra 1 million daily barrels. Even that can only be achieved through the use of gas for secondary recovery and will be hard to sustain for any length of time. But NIOC will be provided with more crude of its own and will decide on destinations, based on political and economic considerations.

62 THE IRANIAN OIL INDUSTRY

NOTES

1. The treatment of the section on nationalization is brief
because there is a wealth of literature on the subject. Some of the
best-known works on the subject are N. S. Fatemi, Oil Diplomacy
(New York: Whittier Books, 1954); A. W. Ford, The Anglo-Iranian
Dispute of 1951-52 (Berkeley, Calif.: University of California Press,
1954); M. Ghassemzadeh, An Analytical Comparison of 1933 and
1954 Agreements (Tehran, 1968), (in Persian); M. Fateh, Fifty Years
of Iranian Oil (Tehran, 1959), (in Persian); Z. Mikdashi, A Financial
Analysis of Middle Eastern Oil Concessions, 1901-1965 (New York:
Praeger Publishers, 1966).

2. It is worth noting that the Russians recognized the need for
a 50-50 profit-sharing agreement before the Western oil companies.

3. For details, see Ghassemzadeh, op. cit., p. 23, and G.
Kirk, "The Middle East 1945-50," Survey of International Affairs,
London, 1954, p. 88.

4. In the 1955-60 period the 50-50 profit-sharing agreement
provided for around 2.2 pounds sterling per ton for Iran.

5. Chairman's statement at the annual meeting of the shareholders, AIOC Annual Report, 1951, p. 15.

6. The treatment of this section is brief because of the wealth
of literature on the subject. For example, see E. T. Penrose, The
Large International Firm in Developing Countries--The International
Petroleum Industry (London: Allen & Unwin, 1968), chaps. 3, 5, 6;
C. Issawi and M. Yeganeh, The Economics of Middle Eastern Oil
(New York: Praeger Publishers, 1962); W. Leeman, The Price of
Middle Eastern Oil (Ithaca, N.Y.: Cornell University Press, 1962);
M. Froozan, "Petroleum Economics," Ph.D. dissertation, Tehran
University, 1963; H. J. Frank, Crude Oil Prices in the Middle East:
A Study in Oligopolistic Price Behavior (New York: Praeger Pub-
lishers, 1966).

7. The term "major oil company" refers to the following seven
oil companies: Standard Oil of New Jersey, Standard Oil of California, Gulf Oil Corporation, Mobil Oil Company, Texaco, Royal Dutch/
Shell, and British Petroleum Company. Compagnie Francaise des
Petroles is sometimes referred to as a major.

8. In 1950 the majors owned over 70 percent of the world's
refining capacity (outside the United States and the Soviet bloc), two-
thirds of privately owned tanker fleets, as well as every important
pipeline. Penrose, op. cit., p. 61.

9. See, for example, Leeman, op. cit.

10. P. H. Frankel, Mattei: Oil and Power Politics (London:
Faber & Faber, 1966), pp. 95-96.

11. Penrose, op. cit., p. 152.

THE CONSORTIUM AGREEMENT 63

12. G. W. Stocking, Middle East Oil (London: Penguin Books, 1971), p. 157.

13. The text of the original agreement can be seen in "Platt's Oilgram News Service," New York, October 5, 1954, and the White Book published by NIOC, History and Text of the Iranian Oil Agreements (Tehran, 1966). (In Persian.)

14. NIOC did not in fact take over the "nonbasic operations" until 1959 and then only partially with financial contributions from the Consortium.

15. Oil Forum, September 1954, p. 307.

16. Reported in Petroleum Intelligence Weekly, April 29, 1974.

17. For the text of the agreement, see Petroleum Intelligence Weekly, Special Supplement, December 25, 1972.

18. Whether the Shah had this arrangement at the back of his mind when he refused to join the participation negotiations, or whether this idea was motivated by the success of the participation program, cannot be factually determined.

19. For the text, see Petroleum Intelligence Weekly, Special Supplement, July 23, 1973.

20. According to Petroleum Press Service (September 1973) the new agreement was to bring Iran an additional sum of $120 million compared to the former Consortium agreement.

21. See Chapter 8.

22. See S. Longrigg, Oil in the Middle East (London: Royal Institute of International Affairs, Oxford University Press, 1961), chap. 5; and Mikdashi, op. cit., chap. 16.

23. Records of the Oil Debates, Iranian Parliament, 1954, quoted in Mikdashi, op. cit., p. 157.

4

THE DEVELOPMENT OF NONCONCESSIONARY OIL CONTRACTS

The Consortium Agreement, which effectively defeated the aims of the 1951 nationalization, caused much controversy among the nations of the Third World about their right to take over and freely exploit their natural resources. This started off a large number of resolutions and counterresolutions in the United Nations General Assembly, the Human Rights Commission, and various other committees. Indeed, the first U.N. resolution was proposed in 1952, the year after Dr. Mossadegh's nationalization attempt and amid the various legal arguments in the International Court of Justice in The Hague. It is not the purpose of this chapter to go into depth about various legal points of the resolutions, as they are well documented.[1] The developing nations of the world were united in proposing a resolution which enhanced the sovereign rights of a nation to nationalize its resources while the industrially developed countries of the world, in particular the United States, the United Kingdom, the Netherlands, and France, with strong backing from New Zealand and South Africa, continuously rejected the resolutions. Indeed, the United States and the United Kingdom had asked for the resolutions to exclude any reference to the "right of self-determination" in the Human Rights Covenants. The arguments dragged on for a decade, and during this period the attitudes of the developed nations had softened in the face of the united opposition of the Third World. The United States and the United Kingdom stated that they were prepared to accept the principle of the sovereign rights of a nation over its natural resources, but they insisted that any particular reference to "nationalization" should be omitted, as this would hamper international cooperation. They further insisted that there should be clear reference to compensation. Indeed they asked for a resolution to include "adequate, prompt, and effective compensation." This was clearly not acceptable to the developing nations. In fact, the Soviet Union and the Eastern European countries insisted that there should be no compensation clause at all in the resolution. Indeed, the Russian representative commented that compensation, in his opinion, could not be paid in accordance with international law, since international law contained no provisions for the compulsory payment of compensation. It was further argued that when national laws allowed compulsory purchase without compensation of the properties of the nationals of a country, that country could not pay compensation to a foreign company in accordance with international laws.

After ten years of bargaining, a compromise was reached in 1962 which was generally accepted--but ironically, the Soviet bloc nations were among the abstainers. The main points of the resolution of December 14, 1962, were

1. Recognition of the principle of permanent sovereignty over natural resources (Par. 1)
2. Nationalization or expropriation should be based on national interest and security (Par. 4)
3. Compensation payable to the nationalized companies should "be appropriate" (Par. 4)
4. Only after all the national jurisdiction of a country which has taken such measures (nationalization) is exhausted, will the controversy, upon the agreement of the sovereign state and other parties concerned, be referred to arbitration and international adjudication (Par. 4).

In 1966 a further resolution was passed by the General Assembly to the effect that "the exploitation of the natural resources of each country shall always be conducted in accordance with its national laws and regulations." The resolution further added that it

> recognizes the right of all countries and in particular, developing countries, to secure and increase their share in the administration of enterprises which are fully or partially operated by foreign capital and to have a greater share in the advantages and profits derived therefrom on an equitable basis, with due regard to the development needs and objectives of the people concerned and the mutually accepted contractual practices, and calls upon such countries from which such capital originates to refrain from any action which would hinder the exercise of that right (Par. 5).

Although it was clear that the passage of such resolutions constituted a great step forward in ascertaining the sovereign rights of
the host states, the resolutions fell short of clearing a large number
of ambiguities. For instance, what would be an "appropriate" compensation, what are the yardsticks for measuring this "appropriate-
ness"? What would be the limit for payment of the compensation?
Presumably the resolutions left the time limit to be determined by
the power struggle between the foreign company and the recipient
state. And most important of all, in what currency should the compensation be paid? Hard currency transfers are banned by national
laws of many developing countries. One may even be tempted to
assume that the importance of all these resolutions is purely academic,
since there are no set precedents for punishment in the cases where
the parties ignore the resolutions. After all, the days of "gunboat
diplomacy" are over and the United Nations has no gunboats anyway.
But it is important to realize that the passage of these resolutions a
decade ago signified a great victory for the developing nations of the
world, although the significance of these resolutions has diminished
greatly today.

In the case of the oil-producing countries, the lesson learned
by the Iranian nationalization proved to be precious. Indeed, they
set about establishing their sovereign rights from the mid-1950s onward. While these countries found themselves at the time powerless
in dealing with the concessionaires (major oil companies), they were
able to devise partnership and service contracts, through which the
state could exercise a fuller control over its oil resources.

THE SIGNIFICANCE OF THE IRANIAN PETROLEUM ACT OF 1957[2]

This act was one of the first well-thought-out and comprehensive petroleum laws of any oil-producing country. It reflected the
tendency to exercise the sovereign rights of Iran over her oil resources, not through hasty actions but by a gradual process of diver-
sifying the sources of crude production.

While NIOC's Constitution remained unchanged under the act,
its functions were broadened and its powers greatly enhanced. The
act recognized NIOC as the owner of all oil resources of the country,
and authorized it "to divide the country including the continental
shelf, but not area covered by the Agreement for the Sale of Oil (The
Consortium Agreement) into districts each of which shall not consist
of more than 80,000 square kilometers" (Article 5). Moreover,
NIOC was asked to ensure that at least one-third of the total exploit-
able area including the continental shelf be set aside as National
Reserves.

NIOC may declare any district, or part thereof, at its discretion as being open to bidding by foreign oil companies. NIOC was empowered to invite other companies to enter into joint agreements for exploration and development of the oil districts. Indeed, the most novel feature of the act was the nature of the joint venture agreements.[3] According to Article 6, NIOC was to hold not less than 30 percent ownership interest in such partnership agreements. In practice, however, NIOC's interest has never been less than 50 percent.

To prevent the repetition of the Consortium Agreement whereby one entity was given control of a large part of the country, Article 7 of the act declared that any joint venture "may hold, in one district and at any one time, not more than 16,000 square kilometers." Furthermore, Article 9G prohibits NIOC from concluding agreements with any company which is involved in petroleum operations in more than five districts at any one time. To ensure speedy development of the petroleum resources, the law requires drilling to start within four years from the date of the agreement, and a specific sum to be spent on exploration. All agreements are to run for 25 years, with provisions for three renewals of five-year terms, and require that each company return to NIOC half of its total area within ten years.

Article 9 of the act sets down the financial arrangements for discovery of oil in the following manner. The partner has to pay NIOC an annual rental to be determined by NIOC; a part of such rental shall be paid as a lump sum in cash upon the conclusion of the agreement, and shall be deducted by annual installments from the payable rental. All rental payments may be included as operating costs until such time as the income tax liability of the operator in any taxation period exceeds or equals the rental payment. Alternatively the act provides that the partner pay all the exploration costs which it would recover if and when oil is discovered in commercial quantities. If no oil is discovered the foreign operator is alone responsible for all the costs. After the discovery of oil in commercial quantities, NIOC supplies its own share of the development costs and shares in the profits of the operation. Only half the oil belongs to the partner, for which the partner will pay 50 percent income tax. The partner may be required to market part or all of NIOC's share of crude. This would give NIOC 75 percent of total profits. Indeed these agreements are known as 75-25 agreements--an inappropriate name, as we shall see later.

Further, Article 2 of the act empowers NIOC to conclude "any agreement which it deems appropriate . . . not inconsistent with the laws of the country." This would practically enable NIOC to enter into any kind of new contract which may seem appropriate to the company. Indeed, it was as a result of Article 2 of the act that NIOC entered into service contracts in 1966.

IRAN'S LEGAL POSITION ON THE OFFSHORE AREAS

Since the 1957 Petroleum Act opened up the offshore areas of the Persian Gulf to exploration, it may be in order to clarify Iran's legal position in the Gulf. The length of the Persian Gulf from the Strait of Hormuz to the farthest point at the head of the Gulf is 805 kilometers, its maximum width 277 kilometers and its minimum width 46.7 kilometers. The present area of the Persian Gulf is about 232,850 square kilometers and Iran occupies the entire northern coast, 805 kilometers in length. The seabed in the Gulf is very uneven, but at its deepest point, near the Strait of Hormuz, the water depth is barely 100 meters.

From the legal point of view the Persian Gulf is divided into the following parts: (a) territorial waters, which include coastal waters. (b) Coastal waters that comprise the sea area between the shoreline and a certain distance, which in the case of Iran is 12 miles. (Territorial and coastal waters are under the sovereignty of the country concerned and are in fact regarded as part of the land of that country.) (c) The seabed adjacent to the coastal waters at the depth of up to 200 meters, is called the continental shelf, and the government has the right of supervising the corresponding sea areas in order to prevent unlawful activities that might endanger the country's security. (d) Waters with a depth of more than 200 meters are referred to as the high seas.

Since the maximum depth of the Persian Gulf is under 100 meters, which is less than the legal depth of 200 meters for the continental shelf, the entire bed of the Gulf may legally be regarded as an extension of the Iranian Plateau. Indeed, the same legal claim may be valid for all the sovereign states of the Gulf.

SIGNOR ENRICO MATTEI'S INFLUENCE ON THE
CONCLUSION OF JOINT-VENTURE CONTRACTS

Enrico Mattei, President of the state-owned Italian Oil Company ENI, had a profound impact on the creation of the joint-venture agreements.[4] Indeed, it may be argued that the Iranian Petroleum Act was shaped in such a way as to legalize the joint venture between NIOC and AGIP (a subsidiary of ENI). It is therefore appropriate to discuss Mattei and ENI.

Mattei was a complex character and an innovator, "a man of many parts. He had in him the makings of a pirate, of an unscrupulous go-getter and of a revivalist preacher."[5] "Nevertheless, a dedicated man whose mission was to secure for Italy a place in the front rank of industrial nations, and also to fight the battle of the

little man against the bosses."[6] After World War II, Mattei realized
that the attempt to discover oil in Italy had failed. He continued the
long-standing arrangement with ENI's main supplier, the Anglo-
Iranian Oil Company, but it was inevitable that Mattei should go
abroad to search for oil. He knew that "there was plenty of low-cost
crude oil available, which was tightly held by a few companies and
that it was sold at a price which was several times its real cost, the
resulting large profit margin being shared equally by the producer
government, and the oil companies."[7] He could see no reason why
Italy should be excluded from this profitable venture, but his attempts to join the elite of international oil were fruitless.

The Iranian nationalization provided a golden opportunity for
Mattei to persuade the major oil companies to allow him some share
of their vast fortunes. Instead of going to Mossadegh's rescue, he
remained totally loyal to the majors, in the hope of a reward in the
form of a percentage interest in the Iranian oil when the dispute was
settled. "Even when the Italian Government acquired 'stolen oil'
from Dr. Mossadegh, through a somewhat obscure firm by the name
of Supor, Mattei would have nothing to do with them, thinking that
such an attitude had earned him gratitude which would be adequately
expressed on a suitable occasion. . . ."[8] The refusal of the major
oil companies to include ENI in the Iranian Consortium was a great
blow to Mattei's loyalty. The refusal was made on the grounds that
independent oil companies could not be included in a major concession. But when, upon the insistence of the State Department, the
majors agreed to include a consortium of nine independent U.S.
companies (the Iricon Group) with a 5 percent interest in the Iranian
Consortium, Mattei's attitude became terribly hostile. He started
a campaign for revenge against the "seven sisters," the aftermath of
which is still visible in the terms of the nonconcessionary contracts.
He created an image of sympathy for the underdog--solidarity between ENI and the exploited nations of the Middle East. He had this
to say to the oil-producing countries of the Middle East:

> The people of Islam are wary of being exploited
> by foreigners. The big oil companies must offer
> them more for their oil than they are getting. I
> not only intend to give them a more generous
> share of the profits, but to make them my part-
> ners in the business of finding and exploiting
> petroleum resources.[9]

Mattei's appeal to the producer nations of the Middle East paid
off. The Iranian Petroleum Act fitted well into Mattei's ideas. It is
interesting to note that the act was ratified on July 31, 1957, in Iran,

and the NIOC-ENI joint-venture agreement was signed a few weeks
later in August 1957. Mattei seemed to have been misunderstood as
being a sincere and businesslike person. In fact he was neither. It
was not through solidarity with the underdog that he initiated the
joint ventures, but rather through a sense of revenge against those
who would not admit him into their select circle. Whatever his motives, he initiated a remarkable change in the power structure of the
international oil system--a change that proved to be beneficial to the
producers. With regard to his economic rationality, Frankel has
written:

> . . . These various tendencies [his attitude toward
> oil producer nations and his quest for low-cost oil]
> were to some extent contradictory, or at least in-
> compatible, and Mattei spent the rest of his life in
> the endeavour to carry on simultaneously, a number
> of policies, which, whenever they were confronted
> with each other, were bound to prove altogether
> troublesome. [10]

Mr. Frankel was at some stage employed as an advisor to the company. On the occasion of his departure from the company, he wrote:
"once he [Mattei] made up his mind about something, he was not
interested any longer in learning what the truth was--he was afraid
it might confuse him."[11]
    The above quotations are not given to reflect the lack of eco-
nomic rationality in proposing joint-venture agreements, but rather
to show that there were overriding political and personal considerations on the part of Mattei in initiating joint ventures.

NIOC-AGIP JOINT VENTURE

    Iran was the first major oil-producing country to sign a joint-
venture agreement.[12] In Iran only the foreign company and NIOC
are signatories to the agreement and the government, as such, is
not a party. NIOC entered into agreement with AGIP, a subsidiary
of Mattei's ENI. The agreement calls for the establishment of the
Iranian-Italian Oil Company, known as SIRIP, within 60 days. SIRIP
is an Iranian corporation in which both partners have equal interests;
each party appoints half the members of the board of directors but
the chairman is an Iranian. The total area under the agreement is
22,700 square kilometers, covering areas in the northern part of the
Persian Gulf lying over the continental shelf, on the eastern slopes
of the Zagros Mountains, and along the coast of the Gulf of Oman.

AGIP agrees to spend at least $6 million within the first four years and $16 million within the next eight years exploring for oil (Article 19). There is no limit to the number of oilfields to be discovered; the $22 million has to be expended regardless of AGIP's production requirements. If after the first four years AGIP decides to suspend its activities, it has to pay NIOC half of the unspent balance from the $22 million.

After the discovery of oil in commercial quantities, each partner pays for half of the total costs. Commercial production was defined to be not less than 100,000 cubic meters per annum (around 86,000 metric tons). SIRIP posts prices in accordance with the prices prevailing in the Persian Gulf (Article 13). The crude may be sold at discounts if approved by the partners. There is no provision for royalty payments, but income tax is payable on AGIP's share of crude at the rate of 50 percent.

Though AGIP is initially responsible for providing the technical personnel, foreign personnel can be employed only when there are no qualified Iranians.

The NIOC-AGIP agreement dealt a blow to the traditional system of concessionary agreements. The trade press, usually reflecting the views of the major oil companies, blasted the agreement as irrational and uneconomic. The facts, however, did not support this allegation. Two months after the SIRIP agreement, NIOC opened District One and within a short period 57 companies from nine countries applied for the NIOC questionnaire. Twenty-two companies returned the questionnaire and 16 of them paid NIOC $2,700 each for a "documentation" on the district. This was in accordance with Article 9.13 of the act, which provided for NIOC to prepare a booklet containing necessary specifications such as geographical and geological situation, plus any other relevant information. Fourteen companies eventually submitted bids. [13]

The successful bidder was Pan American International, a subsidiary of the Standard Oil Company of Indiana. The agreement was signed in June 1958. [14] Although the terms are broadly similar to those of AGIP, the details are different in some important aspects. The main differences can be summarized as follows:

(a) The NIOC-Pan Am agreement called for the creation of a jointly owned company called Iran Pan American Oil Company, IPAC. While SIRIP was an independent commercial enterprise, IPAC was to remain essentially a nonprofit company, acting as an agent for Pan Am and NIOC, jointly controlled by them and producing oil for their account.

(b) There was to be a cash bonus of $25 million, payable by Pan Am to NIOC and recoverable in ten annual installments

after production began (Article 31.5), whereas AGIP did
not have to pay any cash bonus for its contract.

(c) The obligatory expenditure of Pan Am for its exploration
was nearly four times higher than AGIP's. Pan Am had to
spend $82 million compared to AGIP's $22 million over 12
years.

The initial exploration costs and the government tax provisions
are identical in both contracts. Similarly, the foreign partner may
be required to market part or all of the NIOC share on its behalf.

LATER JOINT VENTURES

The IPAC venture proved to be a great success, while SIRIP
had a modest commercial success. Both the ventures started producing commercial quantities of oil in 1961. By 1965 IPAC was producing 2.7 million cubic meters of crude--twice as much as the SIRIP
venture.[15] In April 1963 NIOC announced that it would declare additional offshore areas open for bidding. There was at this stage a
very important problem: lack of information about the structure of
the areas offered for bidding. In the 1957 bidding the 14 interested
companies applied for virtually identical areas lying directly oppo-
site the Neutral Zone (shared between Kuwait and Saudi Arabia)
which had favorable geological features. Thus, apart from the acreage won by Pan Am, the remainder of District One remained uncommitted. It was clear to the NIOC management that there would be no
bidding for the unknown areas. To be able to gather information on
the structure of the uncommitted areas at no cost to itself and to
identify the serious bidders, NIOC devised a plan for the joint financing of a comprehensive marine seismic survey under the auspices of
the Western Geophysical Company over an area of 48,000 square
kilometers. A total of 31 foreign companies, individually or as part
of a group, financed the survey--the most comprehensive study of
its kind at a cost of about $3.5 million.[16]

The survey results, which came out in 1964, were proved later
to have been overoptimistic. At the time, however, they were attractive enough to the participants to submit competitive bids. Of
those who had participated in financing of the survey, only Standard
Oil of New Jersey and Continental Oil Company did not participate in
the bidding.

The maximum acreage to be granted by NIOC depended upon
the type of participation agreement: 8,000 square kilometers when
NIOC had a 50 percent participation, 4,500 square kilometers when
NIOC had less than 50 percent participation, and 2,500 square

kilometers with no NIOC participation at all. Agreements with six groups of companies were signed in 1965 (see Table 4.1). It is interesting to note that Shell had acquired an area on its own in partnership with NIOC. This is particularly important as it reflects the change of heart by the majors and an attempt on their side not to be left behind in this modern type of contract. Indeed, CFP and Socony Mobil were not successful in their bids and their applications were rejected. The successful bidders were a heterogeneous lot. They included one major oil company, three government groups (Italian, French, and Indian) anxious to find a dependable source of supply, and a number of aggressive American independents, the operations of which had hitherto been confined to the United States, but which were interested in obtaining low-cost crude supplies that might enable them to penetrate the international market. The successful bidders agreed to pay cash bonuses aggregating $190 million and to spend more than $129 million over a 12-year period (see Table 4.1). Each partner formed a jointly owned company with NIOC on the lines of IPAC. As stipulated in the Petroleum Act, all partners are responsible for the initial exploration expenditure; costs are shared equally once oil has been discovered in commercial quantities. The profit division in all cases is 75-25, as in the earlier ventures. An additional arrangement was made to guarantee Iran some revenue in cases where marketing would be difficult or delayed. The partner's taxes would have to equal 12.5 percent,at least, of the value of production at posted prices when a commercial oilfield was discovered. There is also a relinquishment clause, 25 percent after five years and 50 percent after ten years, of the total allocated area would have to be returned to NIOC. All the agreements provide for the work to start within six months of signing the contract and that exploration should continue with reasonable speed. The duration of all contracts is to be 25 years, subject to optional renewals of three five-year periods.

1971 OIL AGREEMENTS

On July 27, 1971, NIOC signed three more joint-structure agreements with foreign oil companies. These were the first set of contracts signed after 1965. During this period NIOC gained a great deal of experience with regard to the partnership agreements and was determined to tip the balance of these agreements even more in favor of Iran. The agreements were as follows:

1. NIOC and the Japanese group: The joint company was named Iran Nippon Petroleum Company (INEPCO), and the

TABLE 4.1

Nonconcessionary Joint-Venture Agreements
(millions of dollars)

Company	Short Name	Companies Comprising the Second Party	Effective Date	Area (sq. km.)
Societe Irano-Italienne des Petroles	SIRIP	AGIP S.p.A.	August 27, 1957	22,700
Iran Pan American Oil Company	IPAC	Pan American Petroleum Company	June 5, 1958	14,600
Dashtestan Offshore Petroleum Company	DOPCO	Shell Oil Company	February 13, 1965	6,036
Iranian Offshore Petroleum Company	IROPCO	Tidewater Oil Company Skelly Oil Company Sunray DX Oil Company Kerr McGee Oil Company Industries Inc. Cities Service Company Atlantic-Richfield Corp. Superior Oil Company	February 13, 1965	2,250
Iranian Marine International Oil Company	IMINCO	AGIP S.p.A. Phillips Petroleum Company Oil and Natural Gas Commission	February 13, 1965	7,960
Lavan Petroleum Company	LAPCO	Atlantic-Richfield Corp. Murphy Corp.-Sun Oil Company Union Oil Company	February 13, 1965	8,000
Farsi Petroleum Company	FPC	Bureau de Recherches de Petroles Societe Nationale des Petroles D'Aquitaine Regie Autonome des Petroles (ERAP-ENI)	February 13, 1965	5,759
Persian Gulf Petroleum Company	PEGUPCO[b]	Deutsche Erdol Aktien-geselschaft Deutche Schachtbau und Tiefbohrgesellschaft Gelsenkirchner Berge-werke A.G. Gewerkschaft Elwealth Preussag A.G. Scholven Chemie A.G. Wintershall A.G.	July 15, 1965	5,150
Iran Nippon	INEPCO	Japanese Group: Teijin Ltd., North Sumatran Oil Company Mitsubishi Shojikissha Company	July 27, 1971	8,000
Bushehr Company	BUSHCO[b]	Amerada Hess Corp.	July 27, 1971	3,715
Hormurz Petroleum Company	HOPECO	Mobil Oil Corp.	July 27, 1971	3,506

[a]From commercial production.

[b]PEGUPCO and BUSHCO were dissolved in July and December of 1974 respectively.

Note: See Map of Oil and Gas in Iran.

Cash Bonus	Production Bonus	First 25 Percent Relinquishment	Exploration Period	Minimum Exploration Obligation	Duration, Years
--	--	October 29, 1962	12 years	22	25 years plus 15-year optional[a]
25	--	June 4, 1963	12 years	82	25 years plus 15-year optional
59	28	February 13, 1970	12 years	18	25 years plus 15-year optional
40	--	February 13, 1970	12 years	16	25 years plus 15-year optional
34	10	February 13, 1970	12 years	48	25 years plus 15-year optional
25	6	February 13, 1970	12 years	15	25 years plus 15-year optional
27	2	February 13, 1970	12 years	22	25 years plus 15-year optional
5	5	July 15, 1970	12 years	10	25 years plus 15-year optional
35	10	--	6 years	25	20 years plus 2 five-year optional
5-3/4	6	--	6 years	22	20 years plus 2 five-year optional
3	10	--	6 years	81	20 years plus 2 five-year optional

Source: Iran Oil Mirror, 1968, plus draft of contract for the last three contracts.

allocated area was 8,000 square kilometers of onshore land
in Lurestan. The Japanese group consisted of Teijin Limited, North Sumatra Oil Development Co-operative Company
Limited, Mitsui & Company Limited, and Mitsubishi
Shojikaisha.

2.   NIOC and Amerada Hess Corporation, a U.S. independent
      oil company. The joint-venture was named Bushehr Petro-
      leum Company (BUSHCO). The allocated area was 3,715
      square kilometers of Block I Bushehr offshore land.

3.   NIOC and Mobil Oil Corporation. This was the second
      time one of the seven majors obtained a nonconcessionary
      contract and indeed the first time a major U.S. company
      was involved in joint-venture agreements in Iran. The
      joint-venture company was named Hormuz Petroleum Com-
      pany (HOPECO). The area allocated was 3,500 square
      kilometers of Block II offshore Hormuz Strait.

Departures from 1965 Joint-Venture Agreements

These three new agreements were particularly important as
they constituted important differences from the previous types of
partnership agreements. The main points may be summarized as
detailed below.

(a) Royalty payments or stated payments. All the new agreements provide for the stated payment to the first party (NIOC) of
12.5 to 16 percent of the posted prices applicable to the second
party's share of crude exports. This provision was absent from the
previous joint-venture agreements.[17]

(b) First party's share of crude offtakes. The second party is
bound and obliged to take any quantity of crude made available by the
first party of its share of crude offtake. This provision was voluntary in the previous joint-venture agreements.

(c) Period of the agreements. The new agreements provide
for only six years of exploration and 20 years of exploitation. In the
previous agreements these periods were 12 and 20 years respectively.

(d) Cash bonuses on commercial discovery. The new agreements provide for the payment of bonuses to be related to the cumulative amount of crude production in the assigned area. Thus payment of production bonuses by the second party in the new agreements has been ensured. In the previous agreements, the payment
of production bonuses was based on reaching a given fixed daily crude
production and in practice such levels have never been reached so far.

(e) Board of directors and managing director. In the previous agreements the right of election of the managing director has been vested in the second party. The new agreements provide for an alternation of this right every five years.

(f) Arbitration. Each of the two parties appoint an arbitrator and the two shall appoint a third arbitrator. If there is any dispute the Governor of the Central Bank of Iran, or the Supreme Court of Iran, shall appoint a third arbitrator. In the previous agreements, the final arbitrator is chosen by someone of high standing in a foreign country.[18] This power was transferred to the Iranian side.

(g) Applicable law of the agreement. The new agreements shall be governed by and interpreted according to the laws of Iran. This is in line with the UN General Assembly's resolution of 1966 and would not seek reference to international law.

(h) Loan repayments. Under all agreements NIOC would receive a loan to pay for its share of the development costs. The old agreements had provided for various short repayment periods and did not stipulate any particular range for the chargeable interest rate. According to the new agreements, a fixed time of ten years has been specified for the repayment of loans, beginning from the date of commencement of commercial production. The interest rate under no circumstances would exceed 7 percent.

(i) Renewal of the agreements. The new agreements provide for two additional periods of five years, each renewal by negotiation and revision of the existing terms of the agreements, in the light of the then prevailing circumstances. The old agreements provided three additional periods of five years each, only the third period being negotiable for revision.

(j) Port dues. For the first time the new agreements explicitly provide for the port dues payable to the government of Iran by each tanker calling at the Iranian ports.

(k) NIOC prerogatives. An article specifying the prerogative of NIOC has been included in the new agreements, by which NIOC as the government's representative, will control and supervise the oil operations. Such controls will include the methods, means, and facilities for the crude and product measurements, audit of books and accounts of the joint company, and oil and gas conservation.

(l) Internal consumption. The new agreements clearly specify that,if and when NIOC requires, a certain negotiable portion of the total production should be allocated to NIOC for internal consumption.

(m) Joint use of oil installations. To ensure avoidance of unnecessary investment and to expedite the implementation of the provision of these new agreements, if surplus capacity in pipeline, loading, and other facilities and ancillary services exist in the installation of any oil company operating in Iran, each company, before

embarking on its own developments and installations, must examine
the possibility of utilizing and sharing such available facilities.

(n) <u>Discounts on the posted prices</u>. It has been clearly specified in the new agreements that the income on the total crude exports
--that is, the shares of both first and second parties--will be calculated on the basis of posted prices. Where necessary discounts are
applicable only on the sale of the first party's (NIOC) share of the
crude to the second party. In the old agreement only the share of
the second party was calculated on the basis of posted prices, while
NIOC's share was sold at market prices.

(o) <u>Natural gas</u>. It has been clearly stated in the new agreements that in the event that the second party does not notify the joint
company of its intention to utilize the associated natural gas within
six months, all the available associated gas shall be put at the disposal of NIOC and no charge whatsoever shall be made therefor.

The other provisions in the new agreements are similar to the
1965 partnership agreements.

Clearly the 1971 agreements constitute an important structural
change in the joint-venture contracts. The two principal issues which
were the chief drawbacks of these types of agreement were catered
for in the 1971 agreements--the payment of royalty and the sale of
NIOC's share of crude at posted prices. These two changes have
tipped the balance of profit-sharing greatly in favor of Iran.

On the whole, only 4 of the 11 partnership agreements have
been operational in Iran. They are: SIRIP, IPAC, LAPCO, and
IMINCO. SIRIP, the first joint venture, has been the least successful venture and indeed the smallest producer of oil in Iran. IPAC
has done rather better occupying the second position among the nonconcessionary contractors. The most spectacular success was
achieved by LAPCO (Lavan Petroleum Company) with the discovery
of commercial oil in 1968; LAPCO's production has been increased
nearly tenfold in the span of three years. IMINCO (Iranian Marine
International Oil Company) has also been reasonably successful.
Commercially viable oil was discovered in 1969 and within two years
IMINCO's production rose by around 4.5 times.[19] There is, however,
some feeling of disappointment with regard to the joint ventures.
Four of the six companies (including Shell) which were awarded contracts in 1965 have not yet been successful in striking oil in commercial quantities.

SERVICE CONTRACTS

This type of contract is basically different from the joint-
venture agreements. Although the Petroleum Act of 1957 did not

specifically provide for service contracts, it authorized NIOC to
award any type of contract it sees fit except those which are not in
accordance with the laws of the country (Article 2). The basic principle of the service contracts is that the foreign operator has no ownership rights in Iran at all. It is simply a contractor working for
NIOC and it is remunerated for its services by crude oil. NIOC's
Chairman and Managing Director, Dr. Egbal, at a press conference
on July 27, 1966, characterized this contract as "revolutionary" and
predicted that it would open a new chapter in the history of the Iranian oil industry. The contract is designed both to "increase benefits
accruing to Iran and to develop the economic ties already existing
between the two countries (France and Iran)." The Iranian side was
very boastful about the new contract. Once again Iran was the innovator of a new type of contract and a new concept. This was said to
reflect the maturity and experience of NIOC in its 15 years of activity. Like the NIOC-AGIP joint venture, the new agreement was between two state-owned companies, NIOC and the French Entreprise
de Recherches et d'Activités Pétroliers (ERAP). Indeed, these types
of agreement were thereafter often referred to as "ERAP type agreements." The ERAP group created a subsidiary company named
SOFIRAN, registered in Iran, to take over the activities and respon-
sibilities of ERAP.

Under the agreement, ERAP is required to lend NIOC the necessary funds to cover geological and seismic surveys for an onshore
area of 250,000 and an offshore area of 21,000 square kilometers. The
exploration loan is interest-free, but if oil is discovered ERAP will
lend NIOC further interest-bearing loans to cover the cost of development and production until such time as NIOC's cash flow from this
venture will be sufficient to pay for the expenses (Article 5). If
SOFIRAN finds no oil, ERAP will not recover its exploration loans.

The area under development is considerably smaller than those
covered by the seismic survey. NIOC and ERAP are to select the
most favorable areas: 20,000 square kilometers onshore and 10,000
square kilometers offshore. Within six to seven years the area will
be further reduced to 5,000 and 3,300 square kilometers respectively.

Another important aspect of this contract is the employment of
the concept "national reserves" as stipulated under Article 5 of the
Petroleum Act of 1957. This was set at 50 percent of the discovered
recoverable reserves, but the remaining 50 percent would be developed immediately under NIOC supervision. If SOFIRAN complies
with all its contractual obligations but can find no oil, it can termi-
nate the contract by paying NIOC half the balance of allocated exploration funds. Moreover, all the production equipment, installations, and discovered oil belong to NIOC.

Like the joint-venture agreements, this contract provides for
the minimum exploration of $9 million and $14 million for onshore

and offshore areas respectively (see Table 4.2). The duration of the
contract is 25 years from commercial production of oil, and during
the whole of this period NIOC is obliged to sell ERAP a quantity of the
discovered oil at cost plus 2 percent. The quantity of oil which is to
be sold to ERAP depends on the distance of the field from seaboard. If
the oilfield is located at a distance of 500 kilometers or over from
the shore, SOFIRAN's share will be 45 percent, and if the distance
is 100 kilometers or less, this share will be reduced to 35 percent,
a variation that seems to reflect transport differentials (Article 29,
Sec. I). It is, however, important to emphasize that this 35-45 percent of crude which NIOC is obliged to sell to ERAP is calculated on
the basis of the 50 percent of oil which can be brought into exploita-
tion--that is, SOFIRAN's share would be 35-45 percent of one half
of the field, the other half being set aside as national reserves. This
means that in effect only 17.5 percent to 22.5 percent of the total oil
discovered in the fields is to be sold to ERAP. ERAP will be obliged
to pay 50 percent tax on its profits derived from selling this oil, but
the selling price is to be calculated on the basis of realized prices
and not posted prices (Article 30, Sec. I). From the revenues thus
realized by NIOC, it will repay ERAP for the exploration expenses at
the rate of one-fifteenth of the total expenses annually, or 10 U.S.
cents per barrel of crude bought, whichever is greater (Article 27,
Sec. I). After production has been established NIOC will begin to
repay, with interest, the development loans that ERAP has supplied.
It will repay them in five years either in cash or from the proceeds
of the sale of the crude. According to Article 28, Sec. 5, "each
annuity shall comprise the repayment of loans and interest . . .
computed at the rate of discount of Banque de France, plus 2.5 percent." NIOC can, however, choose to pay in oil. In this case, ERAP,
acting as NIOC's broker, will sell a maximum quantity of 1 million
tons of crude oil each year for the first five years and will retain the
proceeds calculated on the basis of "realized prices minus a brokerage
commission of 2 percent." If this does not yield adequate sums, NIOC
will make good the deficiency in cash. In addition to selling NIOC's
oil to recover its developmental loans ERAP agrees to export to
world markets for NIOC 3 million tons of crude each year during
the first five-year period, and 4 million tons yearly during the second
five-year period. For this oil it will pay NIOC realized prices minus
2 percent brokerage commission.

     A very important clause signifying the political importance of
this contract was included in Article 31: ERAP will only sell oil on
behalf of NIOC if the Iranian government guarantees to use the proceeds to buy French equipment, products, or services as mutually
agreed on by the two governments.

TABLE 4.2

Nonconcessionary Service Contracts

| | Contractor | | |
	SOFIRAN	European Consortium	CONOCO*
Parties to the agreement	ERAP and SOFIRAN	ERAP, ENI, Hispanoil, Petrofina, and OMW	Continental Oil Company
Effective date	12/13/66	6/24/69	4/6/69
Area sq. km.			
Offshore	21,500		
Onshore	254,000	27,260	12,860
Primary exploration period			
Offshore	6 years	8 years	7 years
Onshore	6 years		
Secondary exploration period			
Offshore	3 years	2 years	2 years
Onshore	2 years		
Minimum exploration expenditure obligation			
Offshore	$9 million	$10 million	$12 million
Onshore	$14 million		
Minimum operation obligations			
Offshore	6,000 sq. km. seismic	Three exploratory wells	Three exploratory wells
Onshore	6,000 sq. km. seismic		
Ownership at wellhead	100% NIOC	100% NIOC	100% NIOC
Duration from commercial production	25 years	25 years	25 years
Cash bonus	--	--	$1 million

*In 1974 Continental sold 50 percent of its holdings to Philips Oil Company and 25 percent to Cities Services Petroleum Corporation. The name of the operator was thereafter changed to PHILIRAN.

Note: See Map of Oil and Gas in Iran.

Source: Iran Oil Mirror, 1969 (NIOC Pamphlet).

The basic structure of the service contracts are discussed above and the new contracts awarded later in Iran had the same basic principles. The second service contract was awarded to the Continental Oil Company in April 1969 and the third service contract was awarded in June 1969 to a European consortium. This European consortium is particularly interesting as it consists mainly of state-owned oil companies of Western Europe. The participants are ERAP, AGIP, Hispanoil (Spanish State Oil Company), Petrofina (Belgium), and the Austrian Osterreichische Mineralolwerke (OMW).

1974 EXPLORATION AND DEVELOPMENT AGREEMENTS

In the summer of 1974 the Iranian Parliament passed a new petroleum act as well as a new NIOC constitution. Neither of the above provided for radical changes from those already explained before; instead they consolidated the position of NIOC and fitted in the contract type of agreement not envisaged in the 1957 Act. Furthermore, the law lowers the maximum total period of partnership agreements from 40 to 20 years and that of contract type agreements from 25 to 15 years (Article 3).

The new act lays down regulations governing "exploration and development" agreements, which are basically contract type agreements, the difference being that in this type of contract the foreign contractor is not entitled to any crude at tax paid cost (cost plus income tax) and has to pay market prices. For the purpose of consistency we shall not differentiate between these agreements and the service contracts in the following.

In 1974 NIOC opened up new areas onshore and offshore including the regions surrounding the Consortium's Agreement Area. Although the initial response of the independent and state-owned oil companies was favorable, the very stiff terms required by NIOC resulted in only a few bids.[20] Initially 40 companies are understood to have submitted bids, but by the deadline of June 30, 1974, only CFP, Deminex, and Ultramar had obtained definite contracts. In view of the disappointing response, NIOC negotiated the earlier bids and further agreements were signed with Ashland and AGIP. In all, six contracts were signed with five individual companies, the German Deminex getting two onshore areas around Shiraz and Abadan (see Table 4.3).

The new service contracts constituted a departure from the previous agreements. Under these contracts, as soon as commercial production starts NIOC will assume responsibility for production operations; the oil exploration and development deal ends, the contracting company will be dissolved, and a sales agreement will be

TABLE 4.3

Recent Service Contracts, 1974

Second Party	Date of Signature	Name of NIOC Contractor	Area, sq. km.	Bonus, $ Million	Minimum Expenditures	Period, Years	Discount, Percent
Deminex (Germany)	7/30/74	Deminex Iran Oil Company	6,702	12	32	15	3-3.8
Deminex	7/30/74	Deminex Iran Oil Company	7,810	20	36	15	3-3.8
CFP (France)	7/27/74	Total Iran	8,000	6	40	15	5
Ultramar (U.S.)	8/7/74	Ultramar Iran Oil Company	7,800	4.5	14	15	4-5
AGIP (Italy)	8/25/74	AGIP Iran Petroleum Company	7,150	1	20	15	5
Ashland (U.S.)	8/20/74	Lar Explora- tion Company	7,274	6.25	25	15	3.5-5

Source: National Iranian Oil Company, Details of the New Contracts (Tehran: NIOC Public Relations Office, 1974).

concluded with the foreign operator. The sales agreement covers
35 to 50 percent of the oil discovered at the prevalent market prices
less a 3 to 5 percent discount. The discount is awarded to recom-
pense the exploration company for the financial risk, since it has to
bear all the costs up to commercial production with no compensation
for failure. A further undisclosed discount will be allowed for the
recovery of exploration and development expenditures. Another requirement of these contracts is the payment of cash bonuses of up to
$20 million, and larger minimum exploration expenditures. More-
over, the contract period was reduced to 15 years compared with 25
years in the previous agreements. All the employees of the contracting party will be Iranian, save in very exceptional cases in which the
employment of qualified Iranians is not possible.

The service contracts were the latest in the moves of the producing countries to tighten their control over their natural resources.
Unlike the Consortium Agreement where NIOC had no say in the production and management of the company, joint-venture and service
contracts polarized the decision-making process greatly in the hands
of NIOC management.

With regard to the discovery of oil in commercial quantities,
however, the service contracts have thus far turned out to be a failure. SOFIRAN has found no oil, nor have the others. This of course
does not mean that the contracts have not been progressive, but unless oil is found one cannot really say much about the success or
failure of this type of contract. Indeed, it is quite possible that the
ERAP contract will soon be canceled in Iran, as the provisions of
the agreement provide that unless a commercial discovery is made
within eight years in onshore areas and nine years in offshore areas,
NIOC may cancel the agreement, although for political and prestige
reasons it may not wish to do so. NIOC certainly does not wish to
show its novel concept of contract as being a failure, and would probably allow SOFIRAN to continue its search for oil. But will ERAP
wish to continue after the loss of so much capital in exploration?
This is a question not entirely answerable by economic considerations.
Only time will show what will happen to the ERAP contract.

COMPARISON OF CONCESSIONARY AND
NONCONCESSIONARY OIL CONTRACTS

The details of the concessionary Consortium Agreement were
discussed in Chapter 3, and the main points of the nonconcessionary
joint-venture and service contracts were explained in this chapter.
Let us now attempt to make a comparative evaluation of these various
oil agreements. This comparison is particularly important in view of

the numerous vague statements made by the governments on their profitability. The format of the analysis will be static, and no cash flows are used in the discussion. Since service contracts have not yet provided any actual revenue for the Iranian government, their profitability will be commented on only as a point of academic interest, and indeed, little emphasis will be placed on their profitability calculation in this chapter. On the other hand, we are in a position to make both a theoretical and an actual comparison between the joint-venture and the consortium agreements.

One important point which must be brought out at the outset is that this analysis will have to be confined to a comparison of economically quantifiable factors, therefore political, legal, and economically nonquantifiable factors will have to be excluded from the analytical framework.

The most important differences between a joint-venture agreement and a concessionary agreement are as follows.

(a) The concessionary agreements include an element of royalty at the rate of 12.5 percent of the prevailing posted prices, which were fully expensed by 1971.[21] The majority of joint-venture contracts do not have a royalty clause, although the recent 1971 agreements provided for such royalty payments.

(b) In the concessionary agreements all the oil is sold at posted prices, but in the joint ventures NIOC's share is paid for either at market prices or at halfway prices. Again, in the 1971 joint-venture agreements, provisions were made for NIOC's share to be sold at posted prices, but with an "appropriate discount." The distinction between posted prices and market prices is important, as the posted prices bear no relation to actual realized market prices, the difference being 30-50 cents per barrel in the late 1960s, and up to $2.50 in 1974.

It is a popular fallacy to claim that the concessionary agreements provide for a 50-50 (recently 55-45) profit division, while the joint-venture agreements provide a 75-25 profit division. The governments of producing nations often compare their 75 percent profit take from the partnership agreements to the 50 percent income tax received from the concessionaire to show that the concessionary agreements are less profitable than the joint-venture agreements. This is used to serve a political purpose, by justifying the award of joint-venture contracts and for showing that the governments have been successful in extracting a larger portion of the profits from the new partnership agreements. The basis of this fallacy is that the two kinds of "profits" are not comparable, as each is calculated on a different basis.

Typical Concessionary Agreement (The Consortium)

Posted price for a barrel of Iranian crude in 1971 (on average)	$2.25
Less agreed cost of production (including opportunity cost of capital)	0.20
Less royalty at 12.5 percent of posted prices, fully expensed in 1971	0.28
Net profit	1.77
Government take: 55 percent income tax	0.97
Plus royalty	0.28
Total	1.25
Consortium's take: realized market price	1.80
Less cost of production	0.20
Less payments to Iran	1.25
Total	0.35

The profit split is thus 22-78 in favor of Iran

Typical Joint-Venture Agreement

	Iran's Own Share (dollars)	Iran's Receipts from the Partner (dollars)
Realized export price of a barrel of crude	1.80	2.25
Less cost of production	0.20	0.20
Royalty	--	--
Cost of capital	0.10	--
Income tax at 55 percent	--	1.03
Net profits for Iran	1.50	1.03
Government's take: revenue from own share	1.50	
Revenue from partner (taxes)	1.03	
Revenue from 2 barrels of oil	2.53	
Average from one barrel	1.265	
Foreign partner's take: market price of a barrel of oil	1.80	
Less cost of production	0.20	
Less average payments to Iran for one barrel of oil	1.26	

The profit split in a typical joint-venture agreement is thus
21-79 in favor of Iran. Although the cost of production of the partner-

ship agreements is generally accepted to be around 40 cents per barrel, the cost figures are left at 20 cents to make them comparable to each other. Realized market prices are used in this analysis instead of halfway prices. Halfway prices are calculated as follows:

$$\frac{\text{Posted prices} - \text{cost of production}}{2} = \text{government tax}$$

$$\frac{\text{Posted prices} + \text{cost of production} + \text{government tax}}{2} = \text{halfway prices}$$

In the case of Iran, the half-way price would be $1.74 for partnership agreements. (Some partnership agreements specify halfway prices, while others refer to realized market prices.)

Once we bring the two types of agreement on an equal footing, we will see that the "true profit division" in both cases is similar and that there is no reason to believe one is better than the other, at least in theory.

The previous calculation is based on a theoretical comparison of the profitability of the two types of agreement. Table 4.4 shows what happened in practice.

TABLE 4.4

Per Barrel Disposable Revenue Received by Iran,1968-74
(cents per barrel)

Producer	1968	1969	1970	1971	1972	1973	1974
Consortium	80	85	83	123	133	164	860-960[a]
SIRIP	22	22	22	25	28	24	n.a.
IPAC	40	33	28	51	75	88	n.a.
IMINCO	--	24	31	39	66	106	n.a.
LAPCO	18	29	52	89	100	144	n.a.
NIOC[b]	140	200	174	182	200	348	1,336

[a]Estimated on the basis of expected revenue of $18-20 billion in 1974.

[b]Includes the oil exported by partners, NIOC's exports of partnership crude, and an estimation of the worth of barter trades. Not comparable to the other operators.

Source: Calculated from the figures provided in the Annual Reports of NIOC, 1968-73.

A comparison of the theoretical and the actual benefits to Iran under these two types of agreement leads to the following observations:

(a) The Consortium's per barrel disposable revenue in actual fact is very close to what we might expect from our theoretical calculation for 1971.

(b) The partnership agreements provide a much smaller revenue per barrel than that assumed in our theoretical calculation. Indeed, SIRIP's per barrel disposable income is less than one-fifth of the theoretical amount expected.

(c) The impact of the Tehran Agreement of February 1971, which raised the posted price by 30-35 cents in the Middle East, has been less substantial on the partnership agreements than on the Consortium Agreement. The 50 percent increase in the Consortium's cents per barrel revenue is only matched by LAPCO, in the 1970-71 period.

(d) It is important to realize that the reason for the partnership payments falling behind the Consortium is only partially related to the fact that the government's share of the partnership agreement oil is sold at realized prices and not posted prices. The rise in posted prices has an upward effect on market price, although this upward effect does not necessarily have to be proportional to the rise in posted prices. The main reasons for the bad performance of the joint ventures have been the small economies of scale, offshore production and generally higher production costs, and most important of all, the exemption from the royalty payments, which are linked to the increase in posted prices. (Royalty payments are received from partners equal to the Consortium as of 1973.)

All in all, there is little doubt that the Consortium is the best buyer of Iranian oil in terms of disposable income. However, one may argue that the purely economic benefits might not be the best measurement for the performance of an industry, and that political considerations must enter into it, or that the partnership agreements may have economic effects which are unquantifiable. These arguments may well have their own merits, but they would clearly necessitate present value calculations of cash flows by using an "appropriate" sociopoliticoeconomic discount rate. Obviously, arbitrary "discount rates" will have to be assigned to these cash flows, involving value judgments, and one can come to surprisingly different values by using various discount rates. However, this approach is not practical and economists should direct their attention to the actual situation and draw their conclusions from the facts in their possession. There is no doubt that nonquantifiable economic and political forces are in operation in cases of partnership agreements. Typically they include:

(a) A compromise between national aspirations and the desire to uti-
    lize the natural resources of a developing country
(b) Exercising a tighter control over the natural resources
(c) Better integration of the foreign industry with the domestic economy
(d) Training of Iranians for management positions
(e) Supplying risk capital and foreign exchange
(f) Supplying technical know-how.

In the case of the Iranian oil industry, it can be argued that except for (a) and (b) all the other benefits of the partnership agreements
already accrue to Iran from the Consortium. Indeed, the linkages of
the Consortium with the domestic economy are much stronger than the
partnership ventures, and many more Iranian personnel are trained
and employed by the Consortium compared with the other foreign
partners.

PROFITABILITY IN THE ERAP-TYPE CONTRACTS

The evaluation of the profitability of service contracts has been
a very controversial subject since the ERAP contract was signed in
1966. Various arguments and counterarguments have been used by
economists and government officials to show the division of profits in
this type of contract. As one might expect, different people have
used various assumptions in their analysis to suit their own ends.
Although it is not the purpose of this chapter to go into details, it is
interesting to look briefly at the various methods of calculation.

Dr. M. Egbal, the Managing Director of NIOC, claimed soon
after signing the contract that the Iranian share of profits would
amount to 87-91 percent of the total profits. His argument was based
on a static analysis, which can be summarized as follows.

Assumptions (in cents):
Realized market prices = 150
Cost of production, etc. = 40
Net profit:
   Realized prices - cost of production
         150 - 40 = 110
Income tax
   $$\frac{110}{2} = 55$$
Total cost to ERAP:
   Cost of production + tax + 2 percent of cost of production
         40 + 55 + 0.8 = 95.8
ERAP's profit from one barrel of oil:
         150 - 95.8 = 54.2

Case I: When the contractor is entitled to the purchase of 35 percent of crude discovered, this means that out of 100 barrels of oil discovered, 50 barrels go to the national reserves. From the remaining 50 barrels, 32.5 barrels go to Iran and 17.5 barrels go to the contractor.

NIOC's share of profit $=$ $(50 + 32.5)$ x 110 $=$ 9,075.0
Contractor's share of profit $=$ 17.5 x 54.2 $=$ 948.5
                                                    Total: 10,023.5

Percentage share of NIOC: $\dfrac{9,075 \times 100}{10,023.5}$ $=$ 91 percent

Percentage share of the contractor: $\dfrac{948.5 \times 100}{10,023.5}$ $=$ 9 percent

Case II: When the contractor purchases 45 percent of the oil discovered (excluding the national reserve). This would mean that the contractor would take 22.5 barrels, while NIOC would take $50 + 27.5 = 77.5$ barrels.

Iran's total profit $=$ $(50 + 27.5)$ x 110 $= 8,525$
Contractor's total profit $=$ 22.5 x 54.2 $= \underline{1,219.5}$
                                              9,744.5 cents

Iran's percentage share of total profits
$\dfrac{8,525 \times 100}{9,744.5}$ $= 87.5$ percent

Contractor's percentage share of total profits
$\dfrac{1,219.5 \times 100}{9,744.5}$ $= 12.5$ percent

This static approach came under severe criticism by Dr. T. Stauffer, a leading oil economist from Harvard. In a paper delivered to the Sixth Arab Petroleum Congress in 1967, he produced cash flows showing that the Iranian share of total profits would be 45 percent, and concluded that the ERAP-type contract was less beneficial to NIOC than the Consortium Agreement.[22]

The main points of criticism of Dr. Stauffer were

(a) There is no reason to believe that the "national reserve" (50 percent of the total oil discovered) will be developed and marketed in the same way as the other half. This was implicit in Dr. Egbal's calculations.

(b) There must be an element of opportunity cost of capital in

the region of 15 percent of invested capital introduced into
the calculation.

(c) In the early years, NIOC will have to pay back the loans,
which may be higher than the revenues received. NIOC
may even have to borrow to pay back ERAP loans.

(d) The realized market price of $1.50 assumed by NIOC is
unrealistic. It would be closer to $1.30-$1.40.

According to Dr. Stauffer, if the oil reserves discovered diminish at the rate of 50 percent per year, then the NIOC revenues
from a barrel of crude from the ERAP-type agreement would be equal
to 88 cents, which would nearly equal the Consortium's payment. In
his calculations, Dr. Stauffer ignored the national reserve development of the agreement. Dr. Stauffer's paper was criticized by some
economists,[23] and in 1968 he was invited to come to Tehran to take
part in a debate on his paper. The participants in the debate were
Dr. Mina, an alternate director of NIOC; Dr. Froozan, a high-ranking
NIOC economist; and Mr. Erfani, a former Governor of the Central
Bank of Iran. Later the present author wrote a review of the arguments in the debate.[24] Although the debate was stimulating, no new
points emerged and the differences remained as wide as before.
Dr. Stauffer was criticized for ignoring the discounts on posted price
awarded by OPEC nations to the concessionaires when making his
calculations. Dr. Stauffer claimed that as these discounts were
temporary and would be eliminated, they should not be taken into
account. Indeed, the analytical differences between the participants
were very small indeed, but the differences in using various assumptions by different people were the real cause of the disagreement.

In view of the failure of ERAP to discover any oil in commercial quantities, all these arguments are largely academic and bear
no relation to the actual situation. One may even argue that the idea
of comparing percentages of profit division of one contract with
another may be futile. It is not the percentage points which are of
importance to Iran, but the absolute value of income derived from a
barrel of oil. If Iran's share of the total profit in a contract is only
10 percent but the actual income per barrel received from that contract exceeds the amount received from other contracts, then surely
the former should be preferred to the latter.

In the case of the ERAP contract, all the forecasts are mere
guesses, and no one is in a position to judge the merits of the ERAP
contract for at least five to ten years, after the discovery of oil in
commercial quantities by the company. It is only with the benefit of
hindsight that one can say with certainty if the ERAP contract has
been more profitable or less profitable than the other contracts.

SUMMARY AND CONCLUSION

The last two decades have witnessed the emergence of a new
force in the international petroleum industry. The independent oil
companies were attracted to the Middle East because of the large differences between the cost of crude and the selling price of products.
These independents were basically divided into two groups: the state
oil companies of Western European nations and the independent American and later European and Japanese companies. The state oil companies entered into direct negotiations because their domestic oil
industry was dominated by the international majors; they could see
no reason why they themselves should not enter their own markets or
export the low-cost Middle East oil to other markets. The independents were also concerned with the security of supply. They believed
that in the case of an oil shortage the majors would show their loyalty
to the countries of origin rather than to other consumers. The independent, private oil companies were initially attracted to the Middle
East and North Africa by the prospect of exporting their oil back to
America. Later, when the American import quota policy stopped
their exports to America, they were forced to sell to the European
markets.

In general, all the independents, whether state-owned or private, realized that the only way they could take a share of the cake
was to weaken the dominant position of the international majors and
to gain a foothold in the producing countries. In order to achieve
their objectives they were obliged to concede advantageous terms to
these countries. The trend started with ENI's award of a partnership
contract to Egypt and later Iran--a contract that would cost the host
nation nothing in terms of capital investment until commercially viable
oil was discovered. Early in the 1960s with the weakening of the
dominant position of the majors and the formation of OPEC, offers
of similar joint ventures poured into Iran and other producing countries. The French state oil company, although often ranked as a
major, was quick to seek partnership and service contracts with Iran
as an independent through its subsidiaries. The trend was becoming
clear. The producing countries were intent on exercising their
sovereignty over oil resources and the position of the majors was
weakening. The independents knew that if they began and maintained
good relations with the host governments they would be assured of
supplies in the long run.

After the initial surge of the American independents, the European independents, such as the Belgian "Petrofina," the Austrian
OMV, the European consortium and various German groups, were
offered partnership and service agreements in Iran. The European
independents came mostly from the countries with little or no direct

access to oil supplies. They were concerned both with profits and
assurance of long-term supplies of crude oil. In their attempt to obtain nonconcessionary oil contracts they were strongly supported by
their own governments, whose interests coincided with those of their
oil companies. Moreover, insofar as these contracts did help to
achieve the aims of the producing nations, a political advantage was
secured for the European governments whose independents had
agreed to such contracts.

Any evaluation of oil contracts must take into account the prevailing government policy of the producing country. The Iranian government has an implicit policy of maximizing short-term revenues.
The oil revenues are channeled in increasing amounts to the development budgets of the country. The government is investing heavily
to build a strong economic infrastructure and to encourage industrialization as rapidly as possible. For this the government needs
large amounts of oil revenues. In the 1960s there was an excess
capacity in the world crude production, and it was clear that the
Consortium could not increase its offtake from Iran greatly without
reducing the rate of offtake from other OPEC countries. The reduction in the rate of growth of production in the other producing countries would have been greatly resisted by those countries, and it
could have had undesirable consequences for the concessionaires
(see Chapter 5).

Iran was faced with a situation where she could not expect spectacular increases in the consortium revenues and therefore looked
for other means to increase its much needed oil revenues. It awarded
nonconcessionary contracts which did not live up to its expectations
and effectively brought in smaller revenues compared to those received from the Consortium, in terms of cents per barrel. It is
nevertheless important to realize that within the framework of the
government's policy--that is, the short-term maximization of oil
revenues--the award of any contract, no matter how small its benefits may be, is permissible. Indeed, it may well be argued that even
if the nonconcessionary contracts provide for revenues of even ten
cents per barrel, they should be acceptable, since Iran would not
incur cash payment to the partners and contractors. In this context
the award of nonconcessionary contracts seems to have been fully
justifiable.

NOTES

1. For a detailed analysis see M. Mughraby, Permanent Sovereignty Over Natural Resources (Beirut, Lebanon: Middle East
Research and Publishing Center, 1966); United Nations General

Assembly, Report of the Secretary General, "Permanent Sovereignty Over Natural Resources," Twenty-Fifth Session A/8058, September 1970; M. A. Movahed, Our Oil and its Legal Aspects (Tehran: 1970), in Persian; M. Ganji, Public International Law, Vol. I (Tehran: 1968), in Persian.

2. For a recent translation of the full text, see Iran Oil Journal, issues of September, October, and November 1970.

3. In legal terms these agreements are referred to as "mixed organizations" or "joint structures."

4. The activities of Mattei and his influence on the international oil industry are best documented in P. H. Frankel, Mattei: Oil and Power Politics (London: Faber and Faber, 1966); D. Votaw, The Six Legged Dog, Mattei and ENI--A Study in Power (Berkeley, Calif.: University of California Press, 1964).

5. Frankel, op. cit., p. 25.

6. Ibid., p. 47.

7. Ibid., p. 93.

8. Ibid., p. 95.

9. Quoted in Robert E. Engler, The Politics of Oil: A Study in Private Power and Democratic Direction (New York: Macmillan, 1971), p. 198.

10. Frankel, op. cit., p. 94.

11. Ibid., p. 121.

12. Although Iran was the first major oil producer to sign a joint-venture contract, she was not the first to do so. In February 1957 (six months before the Iranian agreement) ENI and Egyptian concerns formed the Compagnie Orientale des Pétroles D'Égypta-Cope.

13. M. Modir, "The Petroleum Act and the Petroleum Districts," Iran News Letter No. 73 (August 1964).

14. Another joint venture was formed with Sapphire Petroleum, Ltd., a Canadian corporation, in June 1958. This proved to be a commercial failure and was canceled by mutual agreement.

15. For details see Chapter 8.

16. A. M. Mirfakhari, "A Case Study: NIOC's handling of the Marine Seismic Survey," Iran News Letter No. 83 (June 1965). Also, P. Mina and F. Najmabadi, "A New Approach to Full Utilization of Offshore Prospects in the Persian Gulf," ibid., No. 85 (August 1965).

17. Note that the Consortium royalty payments ranged from 7.5 percent in 1966 to 16-2/3 percent in 1974.

18. In the Pan Am agreement, the President of the Swiss Federal Tribunal would appoint a third arbitrator. In the AGIP agreement this power rests with the Chief Justice of Geneva Cantonal Tribunal.

19. See Chapter 8.

20. For details see Iran Oil Journal, August 1974, and Petroleum Economist, September 1974.

21. The royalty rate was increased to 16-2/3 percent of posted
prices in September 1974. The issue of royalty expensing is discussed in Chapter 5.

22. T. R. Stauffer, "The ERAP Agreement: A Study in Marginal Taxation Pricing," Paper presented to the Sixth Arab Petroleum
Congress, Baghdad, March 1967.

23. See K. Sharir, "The Profitability in ERAP Agreement,"
Middle East Economic Survey, May 5, 1967. Dr. Sharir used the
same technique as Dr. Stauffer, but making different assumptions he
concluded Iran's share to be around 65 percent of total profits;
M. Froozan, "Oil Agreements Profitability Comparisons," Iran Oil
Journal (1969); P. Mina, "Changes in the Principle of Oil Contracts,"
Iran Oil News No. 53 (1966), in Persian; M. Froozan, "ERAP Type
versus 50-50 Agreements," Tahquiqat Eqtesadi: The Quarterly
Journal of Economic Research 4, nos. 15-16 (November 1969).

24. T. Stauffer, "The Comparative Evaluation of the Oil Con-
tracts," ibid., Vol. 7 (Summer and Autumn 1970). See also E.
Erfani, "A Comparative Analysis of Oil Agreements," ibid.; M.
Froozan, "ERAP Type Versus 50-50 Agreements: A Further Com-
ment," ibid.; P. Mina, "A Comment on Dr. Stauffer's Paper," ibid.;
F. Fesharaki, "Some Thoughts on the Comparative Evaluation of Oil
Contracts," ibid., Vol. 9 (Winter and Spring 1972).

The purpose of this chapter is to discuss the assertion of the sovereignty of the producing states over their oil resources by collective bargaining through the medium of the Organization of Oil Exporting Countries (OPEC). It is important at the outset to point out that this chapter does not propose to deal with all the activities, functions, and successes of the organization; rather it hopes to deal solely with selected issues which have a direct bearing on the Iranian oil industry and Iranian economy.[1] The selected issues are prices, royalty expensing, proration, the oil price increases of 1971-75, and the chances of OPEC's survival.

THE ISSUE OF CRUDE OIL PRICES

That OPEC was initially created for the purpose of arresting the falling trend of oil prices is well documented in the literature[2] and will not be discussed in detail here. Briefly, up to the end of World War II there was virtually no competitive market (in the economic sense) for crude oil. The major oil companies, which had been operating under the "Achnacarry Agreement" ("As Is" Agreement) of 1928, were able to share the markets in such a way as to balance the supply and demand for crude oil and maintain stable prices. In the 1950s, because of the change in the structure of the industry and entry of independents, the structure of crude prices weakened (see Chapter 3). Insofar as the producing countries were concerned, the importance of crude prices started with the 50-50 profit-sharing agreements. Prior to this, under "tonnage royalty" agreements, the revenues of these countries depended on quantity rather than price. But the 50-50 agreement provided for the direct dependence of the producing countries' revenues on posted prices,

and any decline in posted price would clearly reduce the per barrel
revenues of these countries.

Let us now see what happened to the Iranian crude prices. The
posted price of Persian Gulf oil in June 1957 was $2.04 per barrel.
In February 1959 the price dropped to $1.86 per barrel, and in
August 1960 it was reduced again to $1.78 per barrel. This repre-
sented a fall of 26 cents per barrel. This was the posted price for
oil exported from Mah-Shahr, as the Khark Island facilities had not
yet been built.[3] When they were completed there was an increase of
1 cent per barrel because of the shorter distance and lower loading
and transport costs. Apart from the factors cited above contributing
to the decline of crude oil prices from the Middle East, there was an
element of deliberate action on the part of the oil companies to de-
press the Persian Gulf prices in comparison with the prevailing
prices in the Gulf of Mexico and the Caribbean. Since New York was
chosen to be the "equalization point" for these three major exporting
areas, the price of Persian Gulf crude was kept low enough to be
competitive with the other crude prices at New York, after the freight
cost and import duty was added to it.[4] The Saudi Arabian Oil Minister, Sheikh Abdullah Tariki, argued that the prices of Middle
Eastern crude as posted by the major oil companies during the
decade 1949-59 had been too low in relation to the prices of crude
oil in the Western Hemisphere, and to the prices of products in the
European markets, He concluded that this situation had resulted in
excessive profits for the companies and their affiliates, while it had
deprived the Middle East producing countries of a considerable part
of their rightful share of the profits, which he estimated at $4,743
million.[5] Indeed, the prices of Texan oil (Gulf of Mexico) and
Venezuela (the Caribbean) were nearly 70 percent and 40 percent,
respectively, higher than the Persian Gulf prices of similar API
gravity.[6]

The price reduction of August 1960 provided the turning point.
The producing nations, anticipating further reduction of posted
prices, were forced into making a stand against the falling prices.
In September 1960, the Organization of Petroleum Exporting Countries (OPEC) was formed. Its architects were Dr. Juan Perez
Alfonso, the Venezuelan Minister of Mines and Hydrocarbons, and
Sheikh Abdullah Tariki, the Saudi Arabian Oil Minister. The original
participants were Iran, Iraq, Kuwait, Saudi Arabia, and Venezuela.[7]
The principal aim of the organization at the outset was (a) to stabilize the posted prices, and (b) to restore the pre-1960 posted prices
after the declining trend was arrested. The first objective was at-
tained--that is, the posted prices of crude oil were stabilized for a
decade--but the organization was unable to restore the pre-1960 prices
during the 1960s. The first two years of OPEC's life passed without

any firm resolutions or decisions with regard to oil prices. The first constructive and comprehensive resolution was adopted in June 1962. This was Resolution No. 32 of the Fourth OPEC Conference. This resolution brought out the basic seeds of discontent among the producing nations. It pointed out the dependence of the producing countries on their oil revenues and that the crude oil prices were the determining factor in the welfare of their countries.

The success of this and other resolutions was confined to stabilizing the posted prices and apart from the modest success of reducing/eliminating various discounts and marketing allowances, no real change was made in the structure of posted prices. The stabilization of posted prices did not, however, stabilize the market prices. The realized prices declined and the gap between the posted prices and realized prices widened. Although this bore no direct effect on the producing countries' revenues, it tipped the profit-sharing balance in favor of the producing countries.

THE ISSUE OF ROYALTY EXPENSING

The concept of royalty or "stated payments" had been a part of all oil agreements ever since the first discovery of oil. This concept was used initially in the United States, where oil was owned by private landlords. The companies operating the oilfields were to pay the landowner one-eighth to one-fourth (12.5 to 25 percent) of the oil produced as compensation for the depletion of oil reserves. This royalty was separate from the percentage share of profits paid to the owner of the land or the income tax payable to the U.S. government.

In the 1933 Iranian Agreement, the royalty was fixed at four shillings per ton (with a clause allowing for variation of gold prices), while the profit-sharing principle was maintained at 20 percent for the Iranian side. In this agreement the royalty payments constituted a much larger share of the total payments than the profit-sharing clause. However, in the 1954 Consortium Agreement, and in all other 50-50 agreements, a royalty clause of much less importance than profit sharing was introduced. The royalty was to be at the rate of 12.5 percent of the production of crude oil or its cash equivalent at posted prices.

The disagreement between the OPEC countries and the concessionaires stemmed from the fact that the negotiators of the producing countries had themselves not understood the principle of legal separation of the royalty and income tax payments. Professor Stocking writes:

When I asked Rouhani in 1963, why he had not held
out for the expensing of royalties in the Iranian negotiations with the oil companies, following Iran's
nationalization of the industry a decade earlier, he
said that at the time neither he, nor his fellow negotiators, were aware of the customary practices
in treating oil royalties. Rouhani has learned a
lot about the international oil industry in the inter-
vening years. [8]

In the original 50-50 agreements it was clearly stated that royalty payments were to form a part of the income tax payable to the
producing countries; in other words, the royalties were accepted as
a credit against the 50 percent tax received by the producing countries. Only where the production levels or sales were low was the
royalty payment of 12.5 percent to constitute a floor to the minimum
receipts of the producing governments.

In 1962 Dr. Fuad Rouhani, a brilliant Iranian economist, was
appointed Chairman and Secretary General of OPEC. His task was to
negotiate, on behalf of all the OPEC countries, the expensing of roy-
alties--that is, the separation of royalty payments from the tax pay-
ments--collectively with all the major concessionaires. [9] The first
reaction of the companies was that they did not consider OPEC as a
collective bargaining agency. They also insisted that collective agreements with OPEC might set a precedent for future negotiations and
that such negotiations must be undertaken on an individual basis. It
must be noted that these companies themselves had in the past joined
together to maintain a united front, a privilege they were denying
OPEC. The following example will show the roots of the difference
in an exercise.

Case A		Case B	
Without Expensing of Royalties		With Expensing of Royalties	
Posted price of a barrel		Posted price of a barrel	
of oil in 1970	$1.80	of oil	$1.80
Less production cost	0.20	Less production cost	0.20
Net profits	1.60	Less royalty at 12.5	
Share of the government	0.80	percent	0.225
Share of the company	0.80	Net profits	1.375
Royalties at 12.5 percent		Share of the government	
of posted prices	0.225	(50 percent tax)	0.6875
Share of the government		Total government take	0.9125
through taxes	0.575	Share of the government	
Share of the government		from taxes as a per-	
from taxes as a percent-		centage of profits	50
age of profits	41.8		

In Case A we have a situation with no expensing of royalties. Although the total share of the government is half the total profits, this does not mean that there is an equal profit-sharing arrangement. The companies' payment of 80 cents to the governments included 57.5 cents for income tax and 22.5 cents for royalties. This in effect meant that, excluding royalties, the share of the governments' taxes out of the total net profits would be as follows:

Companies' net profit = posted prices - production cost - royalties
$$\$1.80 - \$0.20 - \$0.225 = \$1.375$$

$$\text{Governments' share} = \frac{0.575}{1.375} \times 100 = 41.8 \text{ percent}$$

What in fact this type of calculation implied was that either the companies were paying no royalties at all or that they paid the royalty correctly but were not paying the 50 percent income tax on net profit as stipulated in the agreements.

Under Case B, the producing countries demanded that royalty be paid separately to the producing countries, but since it is a cost, it may be deducted in reaching a net profit figure. If royalties were fully expensed, the total take of the producing countries would rise by over 11 cents per barrel. It must, however, be emphasized that the agreements made no provisions for the expensing of royalties and legally the concessionaires were not obliged to pay anything more. But the question of legality, on such an important matter involving so many countries, cannot always be interpreted literally, especially when one finds that many of these contracts were awarded under duress, or the host country was ignorant of the situation.

The most comprehensive and intelligent argument for the expensing of royalties can be seen in the 1962 OPEC Resolution, and particularly the Explanatory Memorandum following the Resolution.[10] Article I of the Resolution asked for companies to conform to the recognized principle of the separation of royalty and tax payments, and compensate the producing countries for "the intrinsic value of such petroleum, altogether apart from their obligations falling under the heading of income tax." The Resolution recommended that the countries enter into negotiations with the concessionaires "with a view to working out a formula whereunder royalty payments shall be fixed at a uniform rate and shall not be treated as a credit against income tax liability."

The companies, however, remained unconvinced. They put forward a number of proposals which not only did not aim to even partly satisfy the producers but would also strip them from the right of future negotiations.[11] The arguments and counterarguments went on for two years before a compromise was reached in 1964.

THE SETTLEMENT

Unlike the united stand of the concessionaires, the OPEC members were divided between the militant and the moderate groups.
Iran, Saudi Arabia, Kuwait, and Qatar agreed in principle to accept
an offer made by the concessionaires in 1964. Later Kuwait withdrew but the other four countries signed the agreement, with Iran
being the forerunner because she was convinced that this was the
maximum that the companies were prepared to offer.[12] The settlement permitted the companies, in calculating their income tax obligations, to discount the posted prices by 8.5 percent in 1964, 7.5
percent in 1965, and 6.5 percent in 1966, while the full royalty rate
was to be paid by the companies. This meant that in 1966 the governments would receive 5 cents a barrel more than they would have received under the pre-1964 arrangement. This would be just under half
the 11 cent increase if OPEC's demands were fully met. The settlement provided for consultation in 1966, between the government
and the companies on possible reductions in the future rate of discount. Any agreement reached was to be governed by "the competitive market and economic situation . . . expected at the time of such
consultation" to prevail during future years.[13]

After examining the situation of the international oil industry at
its eleventh conference held in Vienna in April 1966, OPEC recommended that "each member country concerned takes steps towards
the complete elimination of the discount allowance granted to the oil
companies."[14] In accordance with this recommendation, Iran,
Saudi Arabia, and Qatar, and the Secretary General of OPEC started
negotiations with the major oil companies. Negotiations were concerned not only with the elimination of the 6.5 percent tax allowance,
but also with the closely related problem of gravity differential allowances built into the original royalty expensing settlement. In
January 1968 agreement was finally reached with the companies.
Under the agreement, the percentage discounts were to be phased
out in four years, declining to 5.5 percent in 1968, 4.5 percent in
1969, 3.5 percent in 1970, 2 percent in 1971, and were to cease entirely in 1972. The gravity allowance was to be eliminated by 1975.
Libya was not a party to this agreement, but planned to enter separate negotiations on this issue at a later date.

THE QUESTION OF OIL PRORATION

We have already seen that initially the decline in posted prices,
due to the excess supply of crude oil, was the motivation behind the
creation of OPEC. This naturally brings up the question that if excess

supply of oil is responsible for the decline in prices, then the obvious answer would be to prorate or restrict oil production to such a level as to stabilize or even raise crude prices.

The problem, however, is not as simple as it may first appear, and involves many complicated issues. The issue of royalty expensing brought out the first disagreement between the OPEC members, but at least the Persian Gulf states, with the exception of Iraq, presented a united front against the major oil companies. The question of oil proration and the following conflict of interests among the producers even split the Persian Gulf states. It produced disagreement, suspicion, and threats among the producing nations, with far-reaching consequences that threatened the very existence of OPEC. It brought out the inherent contradiction in an organization created to serve nations of conflicting interests. Oil proration was doomed to failure from the beginning, and it was fortunate that it was oil proration and not OPEC itself which disappeared from the scene.

The first and most persistent advocate of oil proration was Dr. Perez Alfonzo of Venezuela. The cost of production of a barrel of oil in Venezuela is 51 cents, compared to the Middle East's average of 15 cents per barrel,[15] while at the same time Venezuelan reserves were believed to be substantially below the Middle East levels. It was clear to Venezuela that excess production was the cause of the weakening structure of international oil prices and that the increasing production would benefit the Middle East at the expense of Venezuela. Indeed in 1960, the year OPEC was formed, Venezuela's production increased by a mere 3 percent over 1959, while total Middle East production increased by 14 percent. Alfonzo was gravely concerned with the declining importance of his country as an oil producer and the subsequent loss of markets and national revenue to the Middle East producers. With this in mind, he set out to persuade the OPEC members to accept a comprehensive system of oil proration.

Among the Middle East producers, Sheikh Abdullah Tariki, the Saudi Arabian Oil Minister, was the official most enthusiastic about production control. Other members agreed to it in principle and it was brought out at the first OPEC conference. The delegates of the member countries, while declaring their intention of stabilizing prices, referred to the regulation of production as one of the means of achieving their objective; and at the second conference, during which the elaboration of an equitable price fixation was envisaged, it was resolved that a study of proration systems should be undertaken. The arguments for proration were entirely based on its impact on prices. An important point which came out of the discussion was that it was not excess production that weakened the price structure, as the level of supply approximately corresponded to the level of demand; the real cause behind the weakening of the prices was the knowledge of excess availability of supply. Accordingly, to influence prices it would be sufficient for OPEC members to make it known to the companies that the available excess supply would not be utilized at the will of the companies. In short, the countries that favored the adoption of a proration policy maintained that if it were possible to apply the quota principle effectively within the United States, it should be equally possible to apply this principle on an international scale.

The arguments against proration were strong. Proration would severely strain the country-company relationships. It would be illegal and the producer would be accused of creating a cartel and interfering with the free operation of market forces. Proration may lead to further intensive research into alternative sources of energy, which would in the long run damage the revenues of producing nations. It was also argued that non-OPEC nations, which were rapidly expanding production, could take advantage of the vacuum and upset the production control program. Even an OPEC member could be tempted to break with the organization and independently come to an arrangement with the companies. The most important argument against proration was the practicality of it: on what basis should the quotas be allocated? Several criteria have been proposed, the principal ones being the volume of proven reserves, the rate of current production and the prospect of its annual growth, the size of investment, the costs of operation, the size of nonoil exports in proportion to oil exports, the size of the population and the degree of economic development--the last two arguments forming the basis of the Iranian demands.

The first positive step toward proration was taken at OPEC's Eighth Conference in Geneva in 1965 when a permanent Economic Commission was established and given the task of drawing up a production control program based on the world demand. Later in 1965 at its Ninth Conference in Tripoli, Resolution IX.61 was adopted which actually provided a quota table for the individual countries.

Country	Allowable Increase (thousands of b/d)	Percent on Previous Year
Iran	304	17.5
Iraq	125	10.0
Kuwait	157	6.0
Saudi Arabia	254	12.0
Qatar	67	32.0
Libya	210	20.0
Indonesia	48	10.0
Venezuela	115	3.3
Average:	160	13.8

Resolution IX. 61 specifically mentioned that these quotas were
to be varied from year to year. The weakness of the resolution was
that there was no planned course of action if either the companies or
the producing countries failed to abide by the quota system. The
quotas were simply proposed to act as a guide to international oil
proration. The main dissenters of quantity restriction were Iran,
Saudi Arabia, and Libya. Iran was unhappy with any kind of proration
because she wanted to secure her former position as the largest Middle East oil producer and because she needed ever-increasing oil
revenues for her development plans. Saudi Arabia could not agree to
a permanent proration because she did not want to lose her lead in
the Middle East, and Libya, because her production was low and she
wanted to expand rapidly. Dr. Reza Fallah, an NIOC Director, made
Iran's position clear in an interview by saying that Iran would never
agree to prorate its production.[16] Similarly, Yamani and Kabasi,
the Oil Ministers of Saudi Arabia and Libya respectively, declared
that they do not agree to a production schedule which may hamper the
growth in production of their crude.[17]

The quota policy was a failure. By January 1966 it was clear
that the first year's production figures for Saudi Arabia and Libya
exceeded those of the allowables, while the rate of offtake in all other
countries, including Iran, was below the proposed output schedule.
It is interesting to note that two of the most important objectives of
the oil proration program--the denial of power to the companies to
penalize a particular country, and the acceptance of output schedules
as maximum allowable by the member countries--had failed. Indeed,
Kuwait and Iraq, which were both engaged in controversies with their
concessionaires, had their production rate cut below the OPEC's
allowables, presumably as a punitive device by the companies--and
OPEC could not, or would not, do anything about it. Also, what
began primarily as a program to curb the rate of increase in output
became one to ensure that each country's output would at least reach
the level stipulated in OPEC schedules.

Iran's Case for Increased Production

NIOC's quest for increased production and its plea for exceptional treatment brought about the first signs of disintegration in
OPEC. NIOC seemed to be departing from the OPEC path, putting
the interests of Iran ahead of the joint OPEC interests.

Iran's demand for increased revenue through higher production
was voiced in the early 1960s when development planning based on
imported technology assumed large scales. There were in fact two
major confrontations between Iran and the Consortium on this issue,
both of which had a direct bearing on Iran's position within OPEC.

In 1966 Iran demanded that oil production be increased by 20 percent to provide the much needed foreign exchange for the economy. The Consortium was uncooperative because of fears that such an increased offtake may have to be exactly matched by a decrease in production of the neighboring countries with dangerous future implications. The result was that Iran issued an ultimatum that unless an agreement was reached within a month, she would take unilateral action (presumably a shutdown) against the Consortium. On December 15, 1966, a compromise was reached. The agreement was in the form of a package deal: the Consortium would return to NIOC 25 percent of the Agreement Area (not stipulated in the 1954 agreement), set minimum output growth schedules of 13-14 percent for the next two years and put at the disposal of NIOC 20 million tons of crude (at low but unspecified prices) over five years for sale to the markets where the Consortium members did not operate (see Chapter 8).

The second major confrontation was in 1968. Iran demanded that the rate of production within OPEC should be linked to the developmental needs of the country and population rather than any other criterion. NIOC stated that in 1967 the production target rate was 12 percent above the previous year's level, falling short by 5.5 percent of the OPEC quota of 17.5 percent. On March 1, 1968, Prime Minister Hoveida, in a speech to the Iranian Parliament, declared his intention of gearing oil output to Iran's needs as envisaged in its development program, saying,

> The oil income figures contained in the 1347 [1968]
> budget reflect neither Iran's true needs nor her
> demands. They merely represent the bare minimum
> required by Iran during the five-year period of the
> Fourth Development Plan [March 21, 1968 to
> March 21, 1973]. We cannot stand by idly while
> our own oil resources are kept unexploited under-
> ground and not utilized for the country's develop-
> ment.[18]

This time NIOC followed a different kind of strategy. It presented to the Consortium a list of Iran's development expenditures, over 63 percent of which came from expected oil revenues. NIOC asked the Consortium to increase the offtake to such levels as to provide NIOC with the specified sum of revenues required for the Development Plan. The Consortium felt unable to comply with NIOC's demand. It pointed out that in keeping with the growth rate for demand of 6 or 7 percent, the Middle East average could not exceed these levels. NIOC's demand required a growth rate of over 20 percent, which was not acceptable to the Consortium. In an endeavor to mobilize and soften Western public opinion in support of Iran's demand for an

increase in oil output sufficient to supply the required development
for funds, Iran sponsored a booklet by David Missen, which presented
the Iranian points of view. The Iranian case was based on the following points.

(a) Population. Iran has the largest population among the producing nations of the Middle East. Indeed, Iran's population is more
than double the total population of other producing countries in the
Middle East. Consequently the rate of oil revenue per capita is lowest in Iran. NIOC argued that the low oil income per head would
justify a larger production for Iran in order to raise the standard of
living of the nation.

(b) Rate of Utilization of Oil Revenues. NIOC claimed that
Iran had the highest rate of utilization of oil revenues for economic
development. Over 75 percent of the oil revenues went to the Plan
Organization, the supreme planning body of Iran, and the rest to
supplement the government budget. Over half the goods imported
every year are capital goods for development projects. Some 12 percent of these imports come from Britain and the rest from other industrially developed countries of the West. It would be in the long-
run interest of the consuming nations to provide Iran with a larger
oil revenue, as this would provide a large market for their exported
goods. Unlike some of its neighbors, Iran has sufficient projects
either socially necessary or economically highly attractive, or both,
to absorb the entire sum of capital which the country's oil industry
is likely to produce. Missen writes,

> Iran has not and would not wish to accumulate sub-
> stantial sums of invested surplus funds, which would
> have to be banked in one or other of the main inter-
> national capital markets. The existence of huge
> surplus oil revenues banked in Britain has inci-
> dentally become a source of great potential insta-
> bility to the British balance of payments position.
> At any time in the past four years it lay within the
> power of the administration of Kuwait, for example,
> to precipitate a devaluation of the pound simply by
> transferring the £700 million worth of surplus
> sterling banked in London into some other currency.
> At no time in the past four hears have reserves of
> this one "statelet" with barely half a million people,
> amounted to less than fifty percent of the available
> foreign exchange reserves of the Bank of England.

He adds,

> A policy of further allocation of production increases
> to countries which cannot possibly absorb the pro-
> ceeds of even their present level of earnings is
> wasteful of scarce funds, less productive of real
> increases in assets, less humanitarian in its prin-
> ciples, less successful in promoting human happi-
> ness. . . .[19]

The statements made by Mr. Missen presumably reflected the official
Iranian view, and it would not be surprising if the other producers in
the region did not take kindly to these statements.

(c) Security of Supply. NIOC officials were quick to point out
that Iran had been the most stable country in the region. Under the
leadership of the Shah, political stability and economic success have
been achieved. In the Arab-Israeli conflict of 1967, Iran was the
only producer in the troubled region to allow not only the continuation
of offtake, but even to increase the rate of production to offset the
loss of oil from other Arab producers. Iran remained politically
neutral in the conflict and was more than pleased with accommodating
the Western consumers. In 1967 the production target for Iran was
12 percent above the previous year, but the Arab-Israeli War, which
did not break out until June, raised production by 21 percent. After
the war the rate of growth of Iranian production dropped to 9 percent
in 1968. For this Iran blamed the lack of foresight by the Consortium
and warned that another clash between Arabs and Israelis would be
unavoidable and Iran would then be once again the savior of the
Western consumers.

Another important message was brought home to the Western
governments: a strong and prosperous Iran, with particularly close
ideological and political ties with Western nations, could play a leading role in making the Persian Gulf a safe waterway for the oil tankers; Iran would also help to slow down the Soviet influence in the
Gulf. Indeed, the departure of the British forces from the Gulf did
not create a politically disastrous "vacuum." The United Kingdom
appeared to have passed over to Iran the task of the Gulf Police.

It was against this background that an agreement with the Consortium was reached on May 14, 1969. The Consortium's planned
target rate for 1969/1970 had fallen £100 million short of the Iranian
demand. The May agreement was a compromise. The Consortium
would meet the full NIOC demand, but the £100 million gap was not

entirely filled by increased production. This was provided in part
by an increase in production of lighter, higher-priced crude and in
part by short-term advances against future revenues.[20] The settlement was to be for one year only (NIOC had submitted the Fourth
Plan requirements for 1969-70, 1970-71, and 1971-72). There was
no actual gearing of production to development needs as Iran had demanded; this was left for future negotiations.

The Consortium settlement brought a sharp reaction from the
other oil producers in the Middle East, who saw the increased Iranian
output as a direct threat to their future production. Kuwait Prime
Minister Al-Sebah, Saudi Oil Minister Yamani, and Iraq's Oil Minister Al-Rifai each made public statements in the press to the effect
that a repetition of such a move by the Consortium will result in
these countries taking punitive action against the major concessionaires.[21]

It was clear to both Iran and the Consortium that a repetition
of such a move, undertaken so openly, would cause a great deal of
trouble for the companies and OPEC. In the event the Iranian production rose by over 18 percent in 1970 and 20 percent in 1971, but
since these increased offtakes seemed to be in line with increased
Saudi Arabian and Kuwaiti offtakes, and since they were thought to be
of the free choice of the Consortium, there was no major trouble
with the other OPEC members.

The End of the Oil Proration Proposals

The oil proration proposals were never officially abandoned,
but they were quietly dropped by the OPEC. The resolutions of the
eleventh conference of OPEC, held in Vienna in April 1966, made no
direct mention of the oil proration and thereafter no reference was
made to it at OPEC conferences.

Why was the oil proration program not successful? We know
that the instability of commodity exports had led producers of coffee,
sugar, tin, and similar commodities to enter into producers' cartels
of "International Commodity Agreements." Some of these cartels
proved to be relatively successful, while others failed to achieve
their objectives. Oil, however, is not strictly comparable to other
raw materials. Because of the lack of substitutes for it in the short
run, and because of the rising demand for products during the 1960s,
the creation of an oil proration program seems to have been theoretically feasible. But the proration program was ill-conceived from
the beginning. It was based on the American proration program without taking into consideration that the Middle East situation was not
suitable for proration. In the United States, because of the problem

of joint reservoirs and the subsequent wastage of gas and loss of
pressure necessary for the lifting of crude, the government had provided statutory regulations for creating a quota system among the
producers. The federal government was able to supervise and enforce proration by legal means. [22] In the Middle East there was no
problem of joint reservoirs, no wastage of gas, and no loss of pressure through over production--and more important, no statutory
means of enforcing the proration. The producing countries were independent states, each trying to gain for their country the maximum
benefit from their natural resources. The inherent conflict between
the producers was based on their different social systems and political and economic objectives. Iran, the only non-Arab state among
the Middle East producing countries, was pressing for higher production and larger revenues to cater for its development expenditure,
while some of the other states, with few outlets for their vast income,
could afford to reduce their offtake in anticipation of higher prices.
The result was complete failure, not only for the oil proration idea
but for OPEC itself. It split the rank of producing nations because it
was the issue that indicated the conflicting interests of the members
most. The abandonment of the proration program pleased all the
parties producers (except Venezuela) and the companies.

THE POST-1971 ERA

The period after 1971 marked a new era in the host country-
oil company relationships within OPEC. The change took place in
two distinct stages: the stage of oil price increases through negotiation with the companies, and the stage of unilateral decisions to increase prices through collective action. Iran's role in developing
the "new attitude" toward indigenous petroleum resources was that of
a leader. The Shah of Iran, with his close ideological ties with the
Western nations, moved further toward an independent national policy
for Iran; indeed, in both stages he asserted himself as the undisputed
leader in the negotiations.

Price Increases Through Negotiations

The 1971 Tehran and Tripoli Agreements, which brought about
the first increase in the posted prices for 11 years, had a twofold
significance. First, they indicated a shift in the balance of power in
the country-company relationship of the oil-producing regions, and
second, they united the ranks of OPEC members which had been split
by the oil proration program. The creation of OPEC in 1960 was a

direct consequence of the falling trend of posted prices; it put a stop
to this declining trend and stabilized prices for a decade.

In 1970 the demand for petroleum throughout the world registered an unusually fast growth. While the growth in the demand for
petroleum products was 6 to 7 percent a year in the last half of the
1960s, in the first six months of 1970 it was 9.4 percent above the
corresponding period in 1969. The unusually cold winter further
contributed to a rise in demand. The Tapline, [23] carrying 500,000
barrels per day, was blown by Palestinian guerrillas in May 1970
and the Syrian government refused to have it repaired. Tanker
freight rates increased as the closure of the Suez Canal made Western Europe more dependent on the low-sulfur Libyan oil. The new
revolutionary government of Libya, which had toppled King Idris in
1969, decided that it was in the public interest to enforce a voluntary
cutback of production so as to avoid overexploitation and consequent
exhaustion of oil resources. The Libyan announcement came four
days after the Tapline was put out of action. Since Libya supplied
about one-quarter of the Western European supplies, the cutback in
production threatened an energy crisis in Western Europe. For the
first time in history the international oil industry was witnessing a
sudden transformation of the buyers' market into a sellers' market,
with energy supplies showing signs of becoming scarce. The United
States was becoming more dependent on the import of Middle East
or Libyan oil, while the antipollution lobby was urging the use of low
sulfur content oil--the same kind of oil that was to be found in large
quantities in Libya. Libya's position was strengthened by the fact
that because of the closure of the Suez Canal and the inadequacy of
the long-haul tanker transport, there was no possibility of tempting
another producer to offset the gap created by the Libyan cutback.
The first company to be ordered to cut back on its production was the
Occidental Petroleum Company. Here Libya showed tactical subtlety
by putting pressure on a company which was in a relatively weak position. Occidental Petroleum, an American independent, had no
other source of foreign crude except Libya, and the company was
thus particularly vulnerable to any production cutback measures.
Later the restriction orders were extended to four other companies,
with the result that, by mid-September 1970, production had been
reduced by nearly 800,000 barrels per day. Libya made it clear to
the companies that an increase of at least 40 cents per barrel would
be the price for lifting the production restrictions. Libya's action
led to an increase of 30 cents per barrel by Occidental, British
Petroleum, and others in September. The prices of all Arab crude
passing to Mediterranean ports were also raised simultaneously.
Libya's action created a golden opportunity for the Persian Gulf producers, which accounted for over 60 percent of world oil exports. It

was agreed that the Persian Gulf states, or the moderates, should make a separate settlement from those of the so-called militants, or the Mediterranean producers, namely Libya and Algeria. The Persian Gulf states' negotiation was taken over personally by the Shah with the approval of the other Gulf states.

On December 9, 1970, OPEC's twenty-first conference opened in Caracas. This conference passed the important Resolution XXI.120, the most comprehensive as well as most effective measure thus far taken by the organization for attaining its fundamental objectives of price restoration. The resolution declared that member countries should take steps to attain the following objectives:

1. To establish 55 percent as the minimum rate of taxation on net income of the oil companies operating in member countries.

2. To eliminate existing disparities in posted or tax-reference prices of the crude oil in the member countries, on the basis of the highest posted price applicable in member countries--taking into consideration differences in gravity and geographical location and any appropriate escalation in the future years.

3. To establish a uniform general increase in the posted or tax-reference prices in all member countries to reflect the general improvement in the conditions of the international petroleum market.

4. To adopt a new system for the adjustment of gravity differential of posted or tax-reference prices on the basis of 15 cents per barrel for 0.1^0 API for crude oil of 40.0^0 API and below, and 20 cents per barrel per 0.1^0 API for crude oil of 40.1^0 API and above.

5. To eliminate completely the allowance granted to oil companies, as from January 1971 (these allowances include discounts from posted price for marketing and the like).

Resolution XXI.120 expressly provided that countries with similar geographical locations, namely the Persian Gulf states, may separately negotiate a deal with the companies. A committee of experts from the Gulf states was to report back to OPEC within 31 days (January 12, 1971) any results of the progress made within the negotiations. Thereafter within 15 days an extraordinary meeting of the organization was to be convened to evaluate the results of the negotiations. If the results were found unsatisfactory, the organization would set out procedures and aim at enforcing unilaterally the organization's objectives through concerted and simultaneous action. In Resolution 122 of the same conference, OPEC noted that any

negotiation must take into account the deteriorating effect of inflation
in industrialized countries and on the purchasing power of member
countries' oil revenues, and resolved that posted prices should be
adjusted to reflect inflationary charges. The resolution pointed out
that the real value of the dollar had fallen by 27 percent since 1960,
while the prices of manufactured articles imported from Western
countries had risen by 35 percent since 1950.

The reaction of the companies was that of initial shock and dis-
belief. To fight OPEC they closed their ranks and continued to resist its unprecedented demands. In their actions they were strongly
backed by their parent governments and by the governments of other
major consuming nations. Indeed, the U.S. government joined with
the governments of Great Britain, France, West Germany, The
Netherlands, Italy, Sweden, and Japan in a united diplomatic front
to back the companies. The U.S. Attorney General, with the approval
of President Nixon, decided to waive the antitrust laws of the United
States to allow the major oil companies to combine against OPEC.
The oil companies asked for negotiations and a "global pact" with all
the members of OPEC. Their insistence on collective bargaining
was a far cry from their attitude in 1962, when they categorically
refused to comply with OPEC's decision to conduct negotiations with
them collectively in the name of all members. The companies'
argument was that regional negotiations would mean that the more
militant members, namely Iraq, Libya, and Algeria, would obtain
better terms. This would lead to "leapfrogging," and the process
could go on indefinitely under the threat of cutting off the supplies.
Negotiations began on January 12, 1971 (the initial OPEC deadline)
and ended on January 21, when the companies agreed to drop their
insistence on a "global pact" and instead agreed to a "regional pact"
if the OPEC assured them that there would be no "leapfrogging"--
that is, that no region would ask for any additional concessions because the companies yielded more in other regions. A final deadline
of February 3, 1971 was set by OPEC. But no results were attained
in the negotiations. The companies had said that the producing
countries' specific demands were too drastic and the deadline was
too stringent. An extraordinary meeting of OPEC was convened on
February 3, and a fourth and final ultimatum was issued in Resolution XXII. 131: "Each member country exporting oil from Gulf terminals shall introduce on February 15 the necessary legal and (or)
legislative measures for the implementation of the objectives embodied in Resolution XXI. 120."

In the event of a failure to reach negotiation by February 15
the member countries, including the other regions, "shall take appropriate measures, including the total embargo on shipments of
crude oil and petroleum products by such company." The biggest

fear of the companies was collective legislation. Since the legislation could not be undone, not only would they have to comply with it, but complying would also set a precedent in the country-company relationship. On February 12 the companies agreed to comply fully with all OPEC's demands.

OPEC's Case for an Increase in Posted Prices

OPEC's case for an increase in posted prices was presented by the Shah, who led the negotiations on behalf of the six Persian Gulf states.[24] The four chief points can be summarized as follows.

(a) The widening of the poverty gap between the rich and the poor nations would require a concession in terms of international trade by the developed nations to the underdeveloped countries, as set out in various UNCTAD (United Nations Conference on Trade and Development) resolutions. What the developing countries--in this case the OPEC members--needed most was not aid but trade. By trade, these countries implied a sort of disguised aid through trade: in effect, by raising the prices of their exports relative to imports they would improve their terms of trade. This was basically a humanitarian appeal, much repeated in UN circles particularly since 1960.

(b) The consumer governments' tax policy was another potentially explosive argument. The large level of taxes levied by the consumer governments on oil products compared to the income tax received by the producing governments is an issue that has been brought up on many occasions. In 1967 the typical breakdown of gasoline in Western Europe was as follows:[25]

Item	Percent
Production costs	2.7
Refining	3.3
Transportation	6.3
Distribution and marketing	26.0
Oil company net profits	6.3
Oil producer's share	7.9
Consumer government's taxes	47.5
Total	100.0

In the United Kingdom the level of taxes is on average over 60 percent.

The OPEC governments were asking for an equal share of the taxes compared to those of the consumer governments. But their

demand raises complex problems. The consumer governments are
entitled to raise their taxes to any level they wish. This is an internal fiscal measure which should be of no concern to foreign governments. The companies are, of course, quite powerless in influencing the domestic tax policy of consumer governments. Also, the
taxes levied on products by the consumer governments are not comparable to the OPEC taxes. The former involves no foreign exchange
component, while the latter is a foreign exchange payment affecting
the balance of payments of the consuming country. Clearly the OPEC
countries should have directed their attack toward the consumer governments and not the companies. OPEC members strongly objected
to the statements made in the Western press that they were responsible for the high prices of petroleum products. They argued that if
the consumer governments felt that the petroleum product prices
were too high they should reduce their own taxes.

(c) The rise in the prices of manufactured goods, coupled with
the standstill in the crude prices, meant that the real purchasing
power of the producer governments declined drastically. The Shah
declared that in the 1960-70 period the cost of living went up by 39.7
percent in the United Kingdom and 24.9 percent in the United States,
while the purchasing power of the U.S. dollar declined by 27 percent. [26]

(d) The increase in the price of petroleum products during the
1960-70 standstill of posted prices was not passed on to the producer
governments. In the United Kingdom, for example, the price of
gasoline was raised by 1 pence per gallon on November 3, 1970 and
by a further 1.5 pence per gallon on December 31, 1970. This meant
an increase of 87.5 cents per barrel, which may be compared to 83
cents per barrel that Iran received from the Consortium in 1970.
Between August and December 1970 the price of oil products rose in
the other Western European countries as follows:

- In nine Western European countries, excluding Italy, 86
  cents per barrel.
- In ten Western European countries, including Italy, 74 cents
  per barrel.
- In the United States the oil companies raised the price of
  crude oil by 25 cents per barrel in 1970, while in Japan the
  prices rose by 22 cents per barrel in October 1970, and
  12 cents per barrel on January 1, 1971. [27]

The OPEC arguments were based on their belief that the price
increases reflected either an increase in taxes of the consumer governments or the companies' intention to raise their profits. However,
it is quite possible that these price increases reflected higher labor
and equipment costs, though it is hard to ascertain the extent to which

the increase in the price of petroleum products reflected the increase
in costs.

The companies' case rested on three arguments. First, the
contents of the concession terms must be respected according to
international law; second, they were not responsible for either the
tax policy of the consumer government or for the rise in the price of
manufactured goods. Finally, the price increase had reflected the
increase in their operating expenses. The companies feared "leapfrogging" by the various OPEC groups, but the Shah assured them
that the Persian Gulf states would not ask for better terms if more
favorable terms were granted to the militant factions of OPEC (such
as Libya and Algeria). Further, OPEC assured the companies
security of supply for the next five years and promised that the governments would not ask for a revision in the structure of the posted
price until 1975. As it turned out this pledge proved to be short-lived.

Terms of the Settlement

The terms of the settlement complied closely with the OPEC
demands as set out in Resolutions 120 and 122. The rate of income
tax was increased by 5 percent to 55 percent; the posted prices of the
Persian Gulf region were raised by a uniform 35 cents per barrel,
which included 2 cents per barrel in settlement of freight disparities;
each of the companies would make a 2.5 percent upward adjustment
to posted prices for inflation on June 1, 1971, and on January 1,
1973 through 1975. In addition, the companies agreed to increase
the crude posted prices by 5 cents per barrel on June 1, 1971, thereafter an increase of 5 cents per barrel was to be added annually on
January 1, 1973 through 1975, to reflect increasing demand for
crude oil during the agreement. From the effective date of the
agreement crude oil was to be posted in the Persian Gulf under a
new system of gravity differentials. For crude oil between 40° and
30° API gravity, each present posted price was to be increased by
5 cents per barrel for each full degree. All the discounts and allowances were to be eliminated.

The Shah, and not OPEC itself, had pressed for a system of
linking the crude prices to an index of international commodity prices.
This was in fact recommended by the United Nations for countries
exporting primary commodities. This objective, however, was not
achieved. The 2.5 percent upward adjustment for inflation was
clearly not enough to stabilize the purchasing power of crude oil, but
in view of the substantial and continuous increases in the posted
prices, this point was not pressed.

In return for their agreement, the concessionaires were assured of stability in the flow of crude oil for five years. No more financial demands were to be made by the Persian Gulf states, and no threat of embargo was to be raised by the producing states.

At the same time, Libya and Algeria declared that the outcome of the Tehran Agreement did not even meet Libya's minimum requirements. An agreement was eventually signed on April 2, 1971. The terms of the Tripoli Agreement were substantially above those of the Tehran Agreement. Although this chapter does not intend to specifically deal with the Tripoli Agreement, we might mention a few of the terms achieved by Libya:

• An increase in posted prices of about 90 cents per barrel
• An increase of 7 cents in posted prices (compared to 5 percent in the Persian Gulf) to reflect the world demand for oil
• An additional surcharge of 9 cents per barrel on top of the 55 percent income tax
• Compulsory reinvestment of the companies' net profit in Libya for further exploration during the five-year period[28]

The impact of the Tehran Agreement was very small indeed on the oil product prices. An increase in the posted price by 35 cents per barrel meant that the governments were receiving 19 cents per barrel more than they did before. This 19 cents reflected the 5 percent increase in income tax and had the effect of increasing the product prices abroad by 5 cents per gallon across the board. Table 5.1 shows the price under the Tehran Agreement through 1975.

Unilateral Price Hikes

From February 1971 to October 1973 the international petroleum industry was relatively calm, despite a few changes. In 1972 the Geneva Agreement provided for an increase of 8.49 percent in the posted prices because of the dollar devaluation. In December 1972 a model participation agreement was signed between Saudi Arabia and Abu Dhabi and the concessionaires for 25 percent equity participation of the governments in the operations.

That the unilateral price rise of October 1973 came just after the start of the Arab-Israeli war was simply a coincidence. The decision to raise the oil prices was taken as early as August 1973 by OPEC. The OPEC members, confident of their newly strengthened alliance and witnessing a surge in demand for oil, decided to test their power by raising the posted prices from $3.00 to $5.12 per barrel. The war, however, induced the Arab producers to stage an

TABLE 5.1

Effect of the Tehran Agreement on the Persian Gulf Prices
(cents per barrel)

API Gravity Degrees	Previous	Price				Producing Countries' Revenues	
		June 1, 1971	January 1, 1973	January 1, 1974	January 1, 1975	As of June 1, 1971	Increase Over Previous
27	147	205.9	216.1	226.5	237.1	113.8	42.8
31	159	218.7	229.2	239.9	250.9	121.6	45.0
34	180	228.5	239.2	250.1	261.4	127.5	40.0
41	195	240.7	251.7	262.9	274.4	134.8	36.7

Source: F. Rouhani, A History of OPEC (New York: Praeger Publishers, 1971), p. 18.

oil embargo against the industrial and "unfriendly" countries in the
hope of bringing pressure on the United States and Israel. The embargo was followed by another unilateral oil price increase in December 1973, when oil prices were increased further to $11.65 per
barrel (see Table 5.2).

TABLE 5.2

Posted Prices and Government Take, 1960-75
(dollars per barrel)

Date	Posted Price[a]	Government Take
1960-70	$1.80	$0.80-$0.85
February 15, 1971	2.18	1.26
1972	2.48	1.44
October 1, 1973	3.01	1.76
October 16, 1973	5.12	3.04
January 1, 1974	11.65	7.00
October 1974	11.65	9.50-10.0
January 1, 1975[b]	10.46	10.12

[a]Light Arabian marker crude of 34 degree API.

[b]Not a posted price, but a single unified price with 22 cents
per barrel allowance for oil company profits and 12 cents per barrel
production cost ($10.46 - [0.22 + 0.12] = $10.12).

Source: Compiled by the author from the trade press, in particular various issues of The Petroleum Economist, Middle East
Economic Survey, and Petroleum Intelligence Weekly.

What were the factors behind the oil price increases of the last
quarter of 1973? First, some increase would have taken place in
any case as a natural outcome of the new OPEC attitude. Second,
the Arab oil embargo and a 30 percent cut in production levels created
an artificial shortage of crude, with panic buying by the smaller oil
companies and the state entities of the industrial countries. In the
auctions held in Iran on December 14 and 20, prices of $17.34 per
barrel were offered for small quantities of Iranian oil.[29] Similar
auctions in Nigeria are reported to have fetched up to $20 per barrel.
This convinced a number of OPEC producers that this was a golden
opportunity to raise prices. Finally, the United States' (and certain
European) backing of Israel persuaded some of the Arab producers
to take punitive action against the industrial countries.

The Shah of Iran was a major advocate of the December price
increase. It is reported that he had asked for the posted price of
$14-$15 per barrel as opposed to the Saudi Arabian demand of $8-$9
per barrel. The Shah took this attitude for three reasons. He knew
that the export horizon of Iran's oil was short and that the prevailing
shortages gave him an opportunity to substantially increase the lot of
oil producers. Furthermore, he was convinced that oil was under-
priced in comparison with the alternative energy sources and such a
price increase would contribute to a more economic usage of this
"noble product."

1974 Price Increases

During 1974, at the time when the market was absorbing the
initial shocks of the oil price hikes in the aftermath of the oil embargo, crude oil prices continued their upward rise. The increases
took place in two ways--open increases and disguised increases.

Open increases took place at various OPEC meetings. At
OPEC's 40th Conference in Quito in June 1974, which met to equalize
the royalty rates in the member countries, it was decided that royalties should rise by 2 percent to 14.5 percent of the posted prices.
At OPEC's 41st Conference in Vienna in September 1974 it was
further decided that royalties be increased to 16.67 percent and the
income tax rate be increased from 55 percent to 65.75 percent.

Disguised price increases came through the development of
three distinct crude prices within OPEC. This new development was
brought about by the increased producer government ownership in
the participation arrangements. Most of the Gulf countries acquired
a 60 percent participation in the ventures (Kuwait took a 100 percent
share in March 1975). This meant that 40 percent of the crude
(equity crude) was sold to the oil companies under the traditional
arrangement at $7 to $7.80 per barrel. Of the remaining 60 percent,
a portion was sold on the free market through auctions or direct
sales by the national oil companies of the producers at around the
same level as posted prices, $11-$12 per barrel; the remainder was
sold to the concessionaires at buy-back prices of 93-95 percent of
postings or around $11 a barrel. The example in Table 5.3 will
clarify the problem.

In view of the confusion in determining the price of crude, the
Shah of Iran proposed a unified price system which will become a
compromise between all prices. In December 1974 Iran's proposal
was adopted by OPEC with only minor alterations. The new price
was to ensure a government take of $10.12 per barrel, with 22 cents

TABLE 5.3

Cost of Crude, 1973-74
(dollars per barrel)

	October 1, 1973[a]	January 1, 1974[a]	July 1, 1974[b]	October 1, 1974[c]
Kuwait (31°)				
Equity crude				
Posted price	2.884	11.545	11.545	11.545
Assumed cost	.060	.060	.060	.060
Royalty	.360	1.443	1.674	1.924
Notional "profit"	2.464	10.042	9.811	9.561
Government tax	1.355	5.523	5.396	6.215
Total government take	1.715	6.966	7.070	8.139
Tax-paid cost	1.775	7.026	7.130	8.199
Buy-back price	2.680	10.850	10.950	?
Average cost of crude[d]	1.957	9.200	8.681	?
Abu Dhabi Murban (39°)				
Equity crude				
Posted price	3.084	12.636	12.636	12.636
Assumed cost	.150	.150	.150	.150
Royalty	.385	1.580	1.832	2.106
Notional "profit"	2.549	10.906	10.654	10.380
Government tax	1.402	5.999	5.860	6.747
Total government take	1.787	7.579	7.692	8.853
Tax-paid cost	1.937	7.729	7.842	9.003
Buy-back price	2.870	11.896	11.980	?
Average cost of crude[e]	2.124	10.128	9.520	?

[a]royalty 12.5 percent; tax 55 percent.

[b]royalty 14.5 percent; tax 55 percent.

[c]royalty 16.66 percent; tax 65 percent.

[d]Assuming a buy-back price of 93 percent of posting for the last quarter of 1973, 94 percent for the first five months of 1974, and 94.85 percent for June and the third quarter of 1974; the volumes of government participation crude bought back as against the companies' equity crude differ from one period to another. Buy-back volumes and prices from October 1974 onward will be negotiated at three-month intervals.

[e]Buy-back volumes and prices were agreed in September and from October 1974 onward will be renegotiated at six-month intervals.

Note: Although Iran was not a party to the participation agreements, she benefited from all the gains through the "balancing margin" principle in its 1973 Agreement.

Source: Petroleum Economist, October 1974, p. 363.

per barrel allowance for oil company profit--a level that may prove
insufficient for profit margins of these companies to sustain operations.[30] The price was then frozen for the first nine months of 1975,
in the hope that during this period a compromise long-term agreement
might be reached with the industrial nations.

OPEC'S CHANCES OF SURVIVAL

The question most frequently raised in recent months by economists, politicians, and media in the West is whether OPEC will
survive and whether the present level of prices can be sustained.
Although this chapter does not propose to investigate this matter in
detail, a number of observations may be made in view of recent
academic and political explorations into the subject.

Professor M. Adelman of the Massachusetts Institute of Technology has long believed that OPEC by virtue of its nature cannot
survive. His argument is based on the case that no "cartel" has been
able to maintain its solidarity for a long period of time in the past.
One or a number of others will inevitably cheat or betray their partners by selling larger quantities of oil in the market and forcing the
prices down. What can be the motive for this action? Adelman's
major argument is based on the excess capacity among the producers.
If the demand for oil falls due to the resistance of the consumers
through quotas or taxes or through economic factors (recession or
price and income elasticities), large non-OPEC oil discoveries are
made and alternative sources of energy developed, then excess
capacity will be generated among OPEC producers. The fall in demand means a fall in revenue at the time of excess capacity, which
will provide sufficient motive for some OPEC members to break the
union. He further urges the U.S. government to help facilitate this
cheating process by various means while maintaining a firm stand
against the producers. Recently, he suggested that the United States
put up oil import quota tickets for competitive bids. These bids
should be transferable and the Treasury should keep the names of the
nominal bidders secret until the quota has taken effect--then "no one
would know who was really behind the nominal bidder--i.e., which
government was willing to chisel on its fellow cartelists and rebate
some of its profits to the U.S. Treasury to keep or expand trade to
this country."[31] He further urges the United States to "seem to mean
business" and to discard the talks of cooperation with OPEC.

Another interesting work on the subject is the Kennedy-
Hauthakker model from Harvard University.[32] Unlike Adelman's un-
scientific approach, the model does have a scientific approach, although the authors have leaped into judgmental conclusions. Kennedy's

model is based on four sectors: supply of crude, refining, transportation, and product demand. The model is explicitly multiregional
and the demand and supply are assumed to depend on their respective
prices. The model, however, is a static one, based on the estimates
of supply and demand elasticities for crude price, how much oil might
be demanded from OPEC in 1980, and at what prices. It concludes that
OPEC members of the Persian Gulf should sell at prices of $3.50-
$5.25 per barrel if they wish to maximize their revenues in 1980.
The model is like an inverted pyramid hinging on the elasticities for
all its validity, and the elasticities are, to say the least, open to
question. Many will find it difficult to believe that at prices of $7 per
barrel the United States will become a net exporter of crude by 1980
or that so much non-OPEC oil will be discovered. Unfortunately,
toward the end of his paper Kennedy jumps into value judgments by
asserting that when the demand for OPEC oil is reduced, these countries will not be able to agree on a prorated production cutback to
absorb the falling revenues and thus will fight among themselves,
resulting in the sale of larger quantities on an individual basis.

   There are also various politicoeconomic factors introduced by
governments and politicians in the West which will in their opinion
lead to a breakup of OPEC. These range from the arguments that
Saudi Arabia, the largest OPEC producers with the largest reserves
and the largest capacity potential, will find it desirable to increase
output and lower prices in order to prevent the development of alternative sources of energy and maintain a cooperative relationship with
the industrial countries. Another scenario points to the eventual
conflict of ideology within OPEC as well as military conflict between
the Arab countries of the Gulf and Iran.

   Such scenarios and conclusions stem from a basic lack of understanding of the people, characteristics, and capabilities of OPEC.
Perhaps it is this misunderstanding of (or the absence of desire to
understand) the basic grievances of OPEC and their legitimate demands that has led to the present situation. Adelman, for instance,
does not distinguish between the association of a number of private
firms and a union of sovereign states with their livelihood dependent
on one product--oil. If indeed OPEC is a cartel, it is in a class by
itself, and to draw an analogy between other cartels and OPEC is a
great mistake. Kennedy was correct in arguing that a drop in demand
for OPEC oil will follow the price increase; indeed, the events of
1974-75 confirm such a reduction. Yet to what extent this was caused
by the world economic recession of 1974-75, and whether the demand
will eventually pick up after the recovery, cannot be factually determined thus far. But to say that OPEC will not be capable of absorbing
a cutback has already been proven wrong. It is true that long-term
interests of Saudi Arabia (in the financial sense) may not be identical

with those of the other producers with lower reserves or capacity potential, but this does not necessarily mean that Saudi Arabia will break OPEC up for financial gain or to please the West. Indeed, the past evidence shows that Saudi Arabia has been a loyal member of OPEC and has taken into consideration the public opinion among the Arab world and its OPEC partners. The greatly publicized auction of Saudi oil scheduled for August 1974, in which the bidding was expected to reduce the prices, did not actually materialize. In November 1974 Saudi Arabia led a number of Arab producers in the Gulf in an action which effectively increased the oil prices. The chances of ideological and military conflict are also negligible. The OPEC producers and particularly the Persian Gulf producers are following an independent national policy, and indeed the Western policy of confrontation has brought them closer together, both politically and economically. Iran's old dispute with Iraq was settled in March 1975, and in April 1975 the Shah visited Saudi Arabia with a reported proposal for a joint Irano-Arab military pact for the security of the Persian Gulf and the Indian Ocean.

OPEC's power clearly stems from its surplus funds (Chapter 7). While proration was unsuccessful in the 1960s because most of the members needed money, the recent events have made it possible for the members to cooperate without the fear of running out of cash. A simplistic example will show the extent of this power. If we assume that 1973 revenues are acceptable to OPEC in the case of a confrontation with the West (after all, in the course of 1973-75 no irreversible need for cash could have developed in these countries) the following can happen. In 1973 OPEC produced nearly 31 million b/d and received a revenue of $23 billion. On the basis of the 1975 government take of $10 per barrel across the board, the output can be reduced to 6.3 million b/d to yield the same income. This means a cut in production of 80 percent in OPEC and a 79 percent drop in world oil exports--a situation that OPEC can accept but the world economy cannot.

It is true that the more developed members such as Iran, Algeria, Indonesia, and Nigeria will suffer some difficulty from a cutback because of their great commitments to internal development, but the shock can be absorbed. Even if the latter three countries are allowed to produce as much as they want, the impact would still be negligible because of their low production level and capacity potential. Some of the smaller states, particularly Abu Dhabi and Omman, ran into financial trouble in early 1975 because of overspending.[33] Indeed, most of the members made very large commitments (perhaps too large) for internal and external expenditures. But to say that this is evidence that OPEC cannot cut back on its spending is to misunderstand the nature of these expenditures: foreign aid, foreign investment in financial and physical assets, loans and trade pacts with the industrial countries, defense, and imports. Such expenditure policies were adopted in the past year or so and can easily be trimmed and readjusted. Indeed, the five richest oil producers--Saudi Arabia, Kuwait, Qatar, Omman, and Abu Dhabi--with 4 percent of the OPEC population and 45 percent of its income in 1974, have the least potential for internal development and will not suffer from any reduced production (Kuwait and Libya have for some years adopted a conservationist policy).

But what is the role of Iran in the OPEC's future? Will the possible shortfall in Iran's foreign exchange requirements (Chapter 7) lead Iran to become a destabilizing force within OPEC, by selling large quantities of crude at lower prices? The answer must be no. First, Iran's oilfields do not have a capacity for large increases in production and second, Iran's low reserves as well as the Shah's commitment to OPEC solidarity will ensure that Iran remains a loyal member of the Organization. Iran will continue to lead OPEC in the future, quite possibly toward higher prices, in line with Western inflation and the costs of the alternative sources of energy.

OPEC's greatest protection against price erosion is that the present prices of $10 or so per barrel are equal or below alternative sources of energy (nuclear and coal energy costs are dependent on environmental regulations). According to some U.S. official estimates, the costs are $6-$7 a barrel for the continental shelf, $13-$15 a barrel for shale oil and $12 a barrel for coal gassification.[34] Indeed, this is why the United States proposed a $9 a barrel as floor price within the International Energy Agency's members to guarantee sufficient return for the investments in the alternative energy sources.

CONSUMER-PRODUCER RELATIONS

The Algiers Summit Conference of the heads of states of OPEC in March 1975--held under a background of sharp drop in demand for OPEC oil, reduced revenues, and some minor price shaving--produced some hope in the West that the organization might yet crack under the pressure of prices and internal conflicts. But instead of confrontation over prices and production quotas, the OPEC Summit preserved its united front and approved a 4,000-word "Solemn Declaration" that could provide a framework for future negotiations with the consumer governments. OPEC made it clear that, under certain conditions, it would be willing to stabilize or even lower the prices slightly, as well as guarantee adequate supplies to the industrial countries. The principal condition is that oil prices be tied to

the world price of other goods. This could be accomplished through
an indexing system based on the prices of manufactured imports, raw
materials, and other commodities and services.

Under the leadership of the United States, the 18-member International Energy Agency (IEA) was created in late 1974.[35] The
original goals of the agency were to coordinate the diverse interests
of the members in the fields of finance, the development of alternative
sources of energy, and the creation of a united consumer body for a
dialogue with OPEC for long-term cooperation.

The preparatory meeting between OPEC and IEA took place in
Paris in April 1975. OPEC was represented by Iran, Saudi Arabia,
Algeria, and Venezuela. At the insistence of OPEC a number of
less developed countries were also invited to the meeting represented
by Brazil, India, and Zaire. Under Algerian leadership OPEC insisted that any long-term agreement on the price of oil must be linked
to the price of other raw materials, and that the purchasing power of
these countries must also be protected. The inclusion of other raw-
material producers in the meeting was a positive step forward for
OPEC policy toward creation of a "new economic order" in which the
other raw-material producers will also share the benefits--perhaps
to the detriment of the rich countries. The IEA members did not
agree and the meeting failed. IEA was basically concerned with
energy as its priority and did not wish to engage in a much broader
agreement, in which OPEC might emerge as the champion of the
underprivileged nations. The end of the meeting was marked by
warnings from Iran and other members that oil prices may well be
increased in September 1975 unless an agreement has been reached
by then.

SUMMARY AND CONCLUSION

OPEC was created initially as an instrument for the stabilization of crude oil prices, very much in line with an international commodity agreement for exporters of raw materials of the Third World.
Because of the very nature of oil the OPEC members were placed in
a strong situation from the beginning. Unlike sugar or coffee which
the Western economies could possibly do without, oil was essential
for the running of the industrial economies of the West.

The first few years of OPEC's life proved to be ineffective, but
the issue of royalty expensing provided a "cause" for unification
among the oil producers. The OPEC members won the agreement
of the companies on the issue of royalty expensing, but had to concede
their claim for the restoration of pre-1960 posted prices.

There was, however, an inherent conflict of interest among the members. Some militant Arab States wanted to use their oil against the Western world in revenge for their support of Israel. The OPEC members were a heterogeneous lot, differing in political ideology, size, population, and degree of economic development. Some members were able to go without the oil revenues for some time, while others were extremely dependent on oil revenues for the running of their economies. The issue of oil proration brought this conflict into the open. Iran was asking for higher production and revenues to pay for its development plans, while Venezuela argued for a unanimous reduction in production rates. The Arab-Israeli War of 1967 and the subsequent stoppage of oil flow from the Arab States to the West did not prevent Iran from stepping up her production to offset the reduced supply. The oil proration program was unofficially abandoned by OPEC, but the ranks of the OPEC remained split.

The 1971 Tehran and Tripoli Agreements were significant in two respects. They closed the ranks of OPEC once again, while revitalizing the issue of restoration of pre-1960 prices, an issue which lay at the heart of the creation of OPEC. Moreover, these agreements provided for a significant shift of power from the companies and consumers to the producers. The international oil market ceased to be a buyer's market and was transformed into a seller's market.

The Tehran Agreement was the first step toward the increasing OPEC power and influence. The agreement was, however, a negotiated agreement, with the Shah personally representing the Persian Gulf nations. The unilateral price increases came at two stages in the last quarter of 1973, and were followed by continuous price increases in 1974, again in most cases Iran being the main advocate of the price increase. As a result, the government per-barrel take of 85 cents in 1970 rose to $10.12 by 1975--a twelvefold increase in the span of five years.

In the 1960s Iran's foreign exchange requirements far exceeded the oil revenues, leading the government to follow a strategy of short-term maximization of revenues through increased production. Iran's pleas for increased revenue and her refusal to join the oil embargoes by following a moderate oil policy toward the West did not pay off. By 1970 Iran had accumulated large external debts and had to resort to ever-increasing borrowing. It had become clear to the Shah that OPEC could not expect an increase in its revenues without a struggle. And in the struggle the Shah emerged as the undisputed leader in the negotiations. Although Iran will no doubt consider its short export horizon in deciding on the future level of prices, the former strategy of increased production for revenue has been abandoned. Iran has declared its readiness to cut back production to maintain the prices and avoid the wasteful usage of oil in the industrial countries.

The idea of indexing oil prices to inflation in the West and to the alternative energy costs were first publicized by the Shah himself. No matter how distasteful these ideas may appear to Western governments, they cannot be ignored. These are solid economic cases which must prevail in the long run. After all, oil prices of $10 a barrel are not, in real terms, higher than the delivered prices of crude in 1950, and the present price is, according to many estimates, below the cost of alternative energy sources.

To smooth out the differences and provide for long-term co-operation a consumer-producer dialogue is badly needed. So far the efforts have been unsuccessful because of OPEC's insistence that other raw materials be also protected against inflation, and because of the general U.S. attitude through dropping hints of military intervention and Kissinger's refusal to accept the validity of a great deal of OPEC's demands. In the short term, OPEC enjoys considerable power because, although demand might fall, alternative sources of energy cannot be developed in the required quantities to threaten OPEC. The technology for some alternative sources which is not yet fully developed cannot be speeded up in proportion to the price increases. For instance, an atomic power plant which needs six to eight years to be constructed cannot be completed in three years if the oil prices are doubled. Thus the transitionary period of a decade or so can be used by OPEC as a bargaining power to obtain long-term cooperation from the industrial countries with a threat of production cutback or price increase.

To believe that OPEC will disintegrate because of excess capacity or internal conflict is wishful thinking. OPEC may well survive the next decade or so. The fruits of solidarity have been too great for OPEC to forget and indeed any new producer in the developing world will join the organization. New suppliers such as the United Kingdom and Norway and the old suppliers such as Canada may not join OPEC, but will doubtless not sell their crude lower than OPEC prices.

NOTES

1. A host of materials is available on OPEC's activities; among them one can name OPEC's publications and resolutions including "OPEC Selected Documents"; F. Rouhani, A History of OPEC (New York: Praeger Publishers, 1971); G. W. Stocking, Middle East Oil (London: Penguin Books, 1971); M. Mughraby, Permanent Sovereignty Over Oil Resources: A Study of Middle East Oil Concessions and Legal Change (Beirut: Middle East Research and Publishing Centre, 1966); M. Movahed, Our Oil and Its Legal Problems (Tehran, 1970), in Persian.

2. See Chapter 3.

3. Shahanshah of Iran on Oil, Tehran Agreement: Background and Prospectives (London: Transorient Books, March 1971), p. 3.

4. Rouhani, op. cit., chap. 12.

5. Ibid., p. 191.

6. Ibid., p. 190.

7. The OPEC membership is now 13. The new members are Qatar (1961), Libya and Indonesia (1962), Abu Dhabi (1967), Algeria (1969), Nigeria (1971), Gabon and Ecuador (1973).

8. Stocking, op. cit., footnote p. 364.

9. Venezuela had already asked for and received an expensing of royalties and was thus not involved in the negotiations. Venezuelan royalty was 16.67 percent of the posted prices. Later Indonesia signed an agreement--structurally different from OPEC's--which did not involve royalty expensing.

10. OPEC Resolution IV, June 1962, and the Explanatory Memorandum. See also "OPEC and the Principle of Negotiations," paper presented at the Fifth Arab Petroleum Congress by OPEC staff, March 16-23, 1965.

11. For a good review, see Rouhani, op. cit.

12. Because of the difference in opinion the negotiations were referred to individual countries. Kuwait signed an agreement in 1967 retroactive to 1964, other countries made their own arrangements.

13. For details of the agreement, see Rouhani, op. cit., pp. 230-33.

14. OPEC Resolution XI.71.

15. C. Tugendhat, Oil: The Biggest Business (London: Eyre & Spottiswoode, 1968), p. 189. (Corresponding Production Costs are: Soviet Union, 80 cents, United States, $1.30, and Indonesia, 82 cents.)

16. Petroleum Intelligence Weekly, May 17, 1966.

17. Middle East Economic Survey, Supplement, February 11, 1966.

18. Middle East Economic Survey, March 8, 1968.

19. D. Missen, Iran: Oil at the Service of a Nation (London: Transorient Books, 1969), p. 23.

20. Middle East Economic Survey, May 16, 1969.

21. Ibid., May 23, 1969, and April 5, 1968.

22. For a good discussion see J. E. Hartshorn, Oil Companies and Governments (London: Faber and Faber, 1967), and S. Schurr, The U.S. Oil Conservation (New York: Resources for Future Inc., 1968).

23. Tapline carries Saudi Arabian oil to the Mediterranean ports.

24. The Shahanshah of Iran on Oil, op. cit. See also the
Press Conference with the Shah in Tehran on January 24, 1971.

25. R. Sanghavi, Iran: Destiny of Oil (London: Transorient
Books, 1971), p. 11.

26. The Shah's Press Conference, op. cit.

27. Ibid.

28. For details of the Tehran and Tripoli Agreements, see
Oil trade press: Petroleum Press Service (now Petroleum Econo-
mist), issues of January, March, April, June, and December 1971,
Petroleum Intelligence Weekly, and Middle East Economic Survey;
Rouhani, op. cit.; OPEC Resolutions XXII 131 and 132.

29. Although some Western newspapers reported that most of
the oil was not in fact lifted, the author was assured by NIOC officials that all the oil was lifted by the buyers.

30. For a good discussion, see Middle East Economic Survey,
Supplement, December 13, 1974.

31. Letter to Wall Street Journal, September 27, 1974.

32. M. Kennedy, Summary of Ph.D. thesis presented to the
Harvard Energy Seminar, November 1974. An edited version of the
thesis, "An Economic Model of the World Oil Market," appeared in
the Bell Journal of Economics and Management Sciences 5, no. 2
(Autumn 1974), pp. 540-77.

33. Reported in Newsweek, March 3, 1975, p. 57.

34. Washington Star, March 7, 1975.

35. IEA members are United States, EEC (except France),
Canada, Japan, Australia, New Zealand, Spain, Switzerland, Sweden,
and Turkey--with Norway as an associate member.

6

OIL AND THE
IRANIAN ECONOMY

The Iranian economy has experienced a spectacular growth rate in the 1960s and 1970s. The growth has come about through sophisticated economic planning and full utilization of the oil revenues. The purpose of this chapter is not to analyze the development of the Iranian economy, which is competently done elsewhere,[1] but rather to show the contribution of the oil industry to the Iranian economy in the past two decades.

This chapter considers the fiscal or indirect influences of the oil revenues through the ordinary and development budgets, as well as the direct influences of the oil industry itself via forward and backward linkages.

IMPACT OF THE OIL REVENUES

The fiscal influence of the oil revenues has a twofold effect on the Iranian economy: first, it provides a source of income for the government to supplement its budgetary expenditures, and second, the oil revenues are channeled through the Plan Organization (now Plan and Budget Organization) for investment in various development projects.

It is necessary at the outset to define the term "oil revenues." Most of the contemporary economic and official publications do not include all the components of oil income, and indeed NIOC's classification differs from those of the Bank Markazi Iran (the Central Bank of Iran) and the Plan organization. NIOC's classification includes the contribution of the domestic oil income to the overall figures as shown in Table 6.1. It can be seen that during 1969-73 the contribution of NIOC's domestic operations (excluding its direct exports)

TABLE 6.1

Iranian Government's Income from the Oil Industry, 1969-73
(millions of rials, NIOC classification)

Details	1969	1971	1973
1. Income tax paid by the Consortium members' trading companies	52,201	109,869	185,800
2. Iranian oil operating companies' income tax	1,033	1,529	11,277
3. "Stated payments" after deduction of general reserve	16,704	29,691	49,680
4. Total receipts from the Consortium	69,938	141,089	246,757
5. Income tax resulting from the SIRIP agreement[a]	160	396	1,822
6. Income tax resulting from the IPAC agreement (share of the second party)	290	710	1,494
7. Income tax resulting from the IPAC agreement (share of NIOC)	636	1,054	1,657
8. Income tax resulting from the LAPCO agreement (share of the second party)	342	2,203	3,993
9. Income tax resulting from the LAPCO agreement (share of NIOC)	633	1,127	3,463
10. Income tax resulting from the IMINCO agreement (share of the second party)	44	258	946
11. Income tax resulting from the IMINCO agreement (share of NIOC)	59	456	1,061
12. NIOC profit tax	1,077	1,948	10,601
13. NIOC dividends	1,000	1,523	8,288
14. Central government and municipal taxes on oil for domestic consumption[b]	8,749	10,823	14,408
15. Employees income tax[b]	834	1,095	1,444
16. Contractors income tax[b]	954	805	1,051
17. Bonus after deducting 2 percent general reserve[c]	75	75	928
18. Total payments, millions of rials	84,791	163,562	297,913
19. Total payments, millions of U.S. dollars	1,116	2,152	4,414

[a]SIRIP's taxes are calculated on the share of both parties.

[b]Collected by NIOC and passed over to the Ministry of Finance and Economic Affairs.

[c]2 percent General Reserve is held by NIOC. The bonus figure for 1973 includes $1 million from the Continental Oil Company, but the larger part of the figure is accounted by royalties (stated payments) which the joint ventures had to pay for the first time from 1973, at the same rate as the Consortium.

Source: National Iranian Oil Company, Annual Reports, 1969-73.

amounted to between 10 to 15 percent of the overall receipts (items
12-16 inclusive). Domestic excise taxes have consistently been the
second largest item after payments by the Consortium--larger than
payments by the joint ventures--amounting to between 5 and 10 percent of the total receipts. Thus the internal operations of the oil industry have on their own been a major source of income for the government.

While such classifications as shown in Table 6.1 are useful for
indicating the order of magnitude of various components, in a strictly
economic sense they may be misleading because they pool foreign
exchange and domestic income together. Foreign exchange is only
received for foreign sales and can only be used to import foreign
goods. Thus foreign and domestic income are not economically comparable, particularly in view of the fact that the quantity and value of
the domestic money may be determined by the government. A more
useful approach would require the construction of a table for the foreign exchange receipts of Iran and the importance of the oil revenues
in total receipts. It can be seen from Table 6.2 that Iran's oil income has increased from $34.4 million in 1954-55 to around $18 billion in 1974-75--an increase of over 500 times in the span of 20 years.
During the 1960s well over half of the total foreign exchange receipts
were from the oil sector. The share of nonoil exports during this
period declined, while foreign borrowing was increasingly used in
the 1968-73 period, accounting for about half the oil revenues and
more than a quarter of the total receipts.

Two important points merit consideration at this stage: first,
the Iranian foreign exchange requirements were far above the oil
revenues and as a result Iran had to borrow substantial sums for her
development expenditures. By March 1973 Iran had accumulated
$5,900 million in foreign debt, compared to $500 million in 1964.
The repayment of long-term foreign debt in the 1971-73 period
amounted to between 11 and 18 percent of the oil receipts.[2] This
situation might provide us with an important motive for Iran to seek
oil price increases in 1973. Second, Iran's ability to borrow internationally is dependent on her oil revenues and oil reserves, which
means that oil revenues have in fact been much more important in
providing Iran's foreign exchange than is shown in Table 6.2.

Thanks to the increasing oil revenues in the Fifth Plan, oil receipts are expected to contribute 86 percent of the total foreign exchange requirements of Iran. Foreign debt will be repaid in large
amounts and only $4,700 million or 4 percent will be borrowed from
abroad ($2,000 million is expected to be the return on Iran's invest-
ment abroad). Foreign borrowing will be confined to the cases when
foreign technology accompanies such loans.

TABLE 6.2

Components of Iran's Foreign Exchange Receipts, 1954-74

Year[a]	Receipts from the Oil Sector[b] Millions of Dollars	Percent	Nonoil Exports[c] Millions of Dollars	Percent	Capital Inflow[d] Millions of Dollars	Percent	Total Millions of Dollars	Percent
1954-55	34.4 (22.5)	15	106.8	48	81.4	37	222.6	100
1956-57	181.0 (240.5)	43	105.8	26	128.9	31	415.7	100
1958-59	344.1 (244.9)	60	141.3	25	89.8	15	575.2	100
1960-61	358.9 (285.2)	60	169.2	28	66.5	12	594.6	100
1962-63	437.2 (342.2)	70	132.6	21	56.5	9	626.3	100
1964-65	555.4 (466.3)	76	146.0	20	30.2	4	731.6	100
1966-67	715.8 (591.5)	65	225.0	20	167.2	15	1,108.0	100
1968-69	958.5 (817.3)	53	366.6	20	496.7	27	1,821.8	100
1970-71	1,268.4 (1,070.8)	54	415.7	18	674.3	28	2,358.4	100
1971-72	2,114.1 (1,885.2)	57	583.6	16	1,013.8	27	3,711.5	100
1972-73	2,536.0 (2,247.1)	58	740.0	17	1,063.7	25	4,339.7	100
1973-74	5,066.6 (4,468.7)	66	1,130.2	15	1,453.1	19	7,649.9	100
1974-75	18,000 (n.a.)	n.a.	n.a.	n.a.	n.a.	n.a.	n.a.	100
Fifth Plan (1973-78)	102,200	86	9,800	8	6,700	6	118,700	100

[a]Iranian calendar starts from March 21, that is, 1974-75 covers the period from March 1974 to March 1975.

[b]The figures in parentheses denote payments from the Consortium; the balance includes payments by all other operators (including NIOC) and all rial purchases by the operators.

[c]Corresponds to all foreign exchange purchases from exporters of goods and services and hence is not identical with actual exports.

[d]Includes utilization of foreign long-term credit and inflow of foreign private loans and capital.

Source: Bank Markazi Iran, Annual Reports and Bulletins, 1963, 1970, and 1973; and the Fifth Development Plan of Iran (revised), Plan and Budget Organization (Tehran, 1975), p. 62.

The Budget

The importance of oil revenues in the development of the Iranian economy may be measured not only by the absolute level of the
inflow but also by its relative importance in the government development budget. The Iranian annual budget has three major components:
(a) the Ordinary or Current Account budget which caters for all current expenditures, the ordinary budget itself being subdivided between the Treasury General and Special Account; (b) the development
or capital budget which shows spending through the Plan Organization;
and (c) the budget of affiliated agencies (public enterprises). Oil receipts are divided between the Treasury General and the Plan Organization. The division of revenues between these two claimants
indicates the government attitude and priorities in utilization of the
oil revenues.
        It can be seen from Table 6.3 that the share of the oil revenues
in development expenditure in Iran increased from 59 percent in 1963
to 80 percent in 1972, signifying the government's determination to
press ahead with industrialization program through imported technology in the country. After the 1973 oil price increases, the share
of the oil revenues allocated to the development planning declined.
This decline does not indicate a change in the government policy, but
rather the inability of the infrastructure of the country to absorb such
large inflows in a short span of time. At the same time, defense
spending increased rapidly and government undertook a series of
social welfare policies through its current budget--policies requiring
imports such as food, clothing, and other basic necessities.
        The Treasury General's revenues have traditionally come from
four major sources: indirect taxes, direct taxes, oil, and deficit
financing. Indirect taxes are by far the largest component of the income, followed by direct taxes and oil. In 1964 oil receipts accounted
for 31 percent of the receipts of the Treasury General, but from the
mid-1960s this ratio fell to 16-20 percent of the disbursements.
After 1973 oil income was increasingly used by the Treasury General
as explained above.[3]

Development Planning

Development planning started in Iran in 1949 after the parliament passed the Plan Organization Act, establishing a machinery for
planning and implementing the development projects in Iran. From
1949 to March 1973, four development plans were executed in Iran.
The First Seven Year Plan coincided with the disruptions of the
nationalization of the oil industry in 1951 and failed with only 20

TABLE 6.3

Division of the Oil Revenues Between the Treasury General
and the Plan Organization, 1963-73

Year	Oil Revenues, Billions of Rials	Share of Treasury General Billions of Rials	Percent of Total	Share of the Plan Organization Billions of Rials	Percent of Total
1963-64	27.7	11.4	41.2	16.3	58.8
1964-65	36.4	14.1	38.7	22.3	61.3
1965-66	50.0[a]	12.4	24.8	37.6	75.2
1966-67	47.4	13.3	28.1	34.1	71.9
1967-68	54.0	14.5	26.8	39.5	73.2
1968-69	61.8	15.0[c]	24.3	46.8[d]	75.7
1969-70	70.0[b]	14.7	21.0	55.4	79.0
1970-71	83.8	17.6	21.0	66.2	79.0
1971-72	150.3[a]	34.0	22.6	116.3	77.4
1972-73	178.2	36.6	20.5	141.6	79.5
1973-74	311.2	91.2	29.0	220.0[e]	71.0
Fifth Plan (1973-78)	6,732	2,125	32.0	4,607[f]	68.0

[a]Includes oil bonuses of 10.5 billions of rials and 3.5 billions of rials in 1965 and 1971.

[b]Excludes 6.3 billions of rials advance payment by the Consortium.

[c]Includes 2 billions of rials transfer from the Plan Organization oil revenues for the implementation of the new civil code.

[d]Excludes 2 billions of rials mentioned in (c).

[e]Estimated.

[f]Estimated by assuming that 100 percent of Plan Organization's requirements in the Fifth Plan will come from oil.

Source: Bank Markazi Iran, Annual Report and Balance Sheet, 1968-73 and the Fifth Development Plan of Iran (revised), Plan and Budget Organization (Tehran, 1975).

percent of the funds actually disbursed. The Second Seven Year Plan
also met with an economic recession in the early 1960s and a severe
drought in agriculture. The first two plans were more in the nature
of impact projects and financial allocations than plans, with relatively
little success. The third and fourth Five Year Plans were much more
sophisticated and ambitious and brought about a high rate of economic
growth in Iran as well as the transformation of the Iranian society
from an agrarian economy into a modern structure capable of industrial development. [4]

Table 6.4 shows how dependent the development projects are
on the oil revenues, and to what uses the funds have been put. Over
the whole period, the share of agriculture has steadily declined from
25 to 15 percent, while investment in industry and mines has increased to 19 percent in the Fourth Plan compared to 7 percent in
the Third Plan. In total the largest share of investment has gone to
infrastructure: transport and communication and other items such as
health, education, manpower training, housing and construction, and
collection of statistics. This was necessary in view of the fact that
until the 1960s very little infrastructure was available in Iran, and
the industrialization process could not get on its way before such
improvements were undertaken.

Structure of Imports

So far we have discussed the foreign exchange receipts of Iran
and their allocation to various projects. In order to better understand the impact of such investments we need to know how the available foreign exchange was used and what type of goods were imported.
Table 6.5 shows the structure of the imports in the past decade. The
largest component of the imports is intermediate goods, followed by
capital goods. Consumer goods, contrary to the popular belief, are
the smallest import item, amounting to 15 percent of the total in
1973-74. Industries and mines have consistently received the lion's
share of the imports, more than 65 percent in most years. Agricultural imports have had a minimum share, less than one-tenth of 1
percent of the total. The composition of the imports shows the government strategy for utilization of the oil revenues in order to indus-
trialize the economy.

The Fifth Development Plan

The Fifth Plan is to run from March 1973 to March 1978. The
original plan drawn up in 1973 envisaged total investment of $32 bil-
lion, of which $17 billion was expected to come from the oil revenues

TABLE 6.4

Revenues and Disbursements of the Plan Organization
(billions of rials)

	First Plan[a] (1949-55)		Second Plan (1956-62)		Third Plan (1963-67)		Fourth Plan (1968-72)	
	Amount	Percent	Amount	Percent	Amount	Percent	Amount	Percent
Revenues								
Oil	7.8	37	61.0	73	153.0	66	385.0	63
Other[b]	13.2	63	22.2	27	79.0	34	225.0	37
Total	21.0	100	83.2	100	232.0	100	610	100
Disbursements								
Agriculture	5.25	25	17.4	21	47.3	20	92.5	15
Transport and communication	5.75	28	27.3	33	53.8	23	113.8	19
Fuel and power	1.0	5	c	—	32.0	14	94.8	16
Industry and mines	3.0	14	7.0	8	17.1	7	113.1	19
Social services[d]	6.0	28	9.3	11	33.3	15	38.9	6
Other projects[e]	—	—	9.0	11	21.1	9	53.7	9
Total development	21.0	100	80.0	84	204.6	88	506.8	84
Nondevelopment[f]	—	—	13.2	16	27.4	12	103.2	16
Total	21.0	100	83.2	100	232.0	100	610.0	100

[a]Only 20 percent of the planned expenditure was actually disbursed.
[b]Includes foreign and domestic borrowing.
[c]Included in industry and mines.
[d]Includes health, education, and social welfare.
[e]Includes housing and construction, municipal and regional development and statistics.
[f]Includes repayment of foreign and domestic loans and administrative expenditures.

Source: Plan Organization, the Report on the Second Seven Year Plan; the Report on the Performance of the Third Five Year Plan, 1963 and 1968. Economist Intelligence Unit, Annual Supplement, "Iran," 1970 and data from the World Bank.

TABLE 6.5

Composition of Iran's Imports, 1964–73
(millions of U.S. dollars)

	1963	1964	1965	1966	1967	1968	1969	1970	1971	1972	1973
Intermediate goods	285.1	408.0	518.2	558.2	711.0	856.5	987.3	1,068.5	1,336.3	1,596.2	2,273.7
Industries and mines	221.9	317.9	410.1	444.5	545.3	641.7	737.4	845.0	1,110.9	1,265.8	1,912.0
Construction	34.8	58.8	69.2	74.7	120.8	147.0	152.7	145.8	138.5	204.3	237.8
Services	22.1	24.7	30.7	31.3	30.9	52.1	64.8	52.7	57.8	97.4	76.3
Agriculture and livestock breeding	6.3	6.6	8.2	7.7	14.0	15.7	32.4	25.0	29.1	28.7	47.6
Capital goods	104.4	162.4	223.0	260.7	329.3	376.3	387.2	391.0	482.9	642.6	906.0
Industries and mines	54.0	72.8	132.1	160.1	230.2	239.1	316.2	263.7	316.6	411.9	560.3
Services	26.7	47.7	55.8	63.9	71.5	103.8	30.9	91.2	132.7	168.4	273.0
Agriculture	23.7	41.9	35.1	36.7	27.6	33.4	40.1	36.1	33.6	62.3	72.7
Consumer goods	124.0	171.9	157.2	144.8	150.0	156.4	168.2	217.1	241.7	331.6	557.4
Total	513.5	742.3	808.4	963.7	1,190.3	1,389.2	1,542.7	1,676.6	2,060.9	2,570.4	3,737.1

Source: Foreign Trade Statistics of Iran.

(total oil revenues were estimated at $22 billion). After the Iranian
takeover of the Consortium in the summer of 1973, the government
announced that the plan would be revised. Before the planners could
draw up new estimates the two price increases of the last quarter of
1973 came into effect, and the fully revised plan was presented to the
Parliament in December 1974. The plan is on its own larger than all
the previous plans put together. With the oil revenues in 1974 alone
exceeding the total allocated to the original Fifth Plan, the government announced anticipated investments of $69.6 billion in this period.
Of the total, $46.2 billion will be in the public sector and $23.4 billion in the private sector. The growth of investment in the public
sector is expected to rise from 14.1 percent per year in the Fourth
Plan to 38.1 percent per year in the Fifth Plan.

 Unlike the previous plans, the new plan does not divide the total
receipts into the needs of the Treasury General and the Plan Organization; rather a table of the total income is provided. Of the total receipts of $122.8 billion, $69.6 billion will be expended by the Plan
Organization (see Table A.1).

 The anticipated balance of payments shows receipts of $102.2
billion from the oil revenues which is expected to finance nearly all
of the Plan Organization's requirements. Imports of goods and service is projected to grow at an astonishing rate of 60 percent to a
total of $93.4 billion in the plan period. Debt repayment will amount
to $6.5 billion and is expected to rid Iran from most of her long-
accumulated debts. An interesting feature of the new plan is an anticipated investment of $11 billion abroad with an expected return of
$2 billion in this period (Table 6.6).

The Overall Impact

 During the 1960s and early 1970s Iran experienced a remark-
ably high rate of economic growth in current prices, constant prices,
and per capita income. The Third Plan target of 6 percent was surpassed by 3 percent and the Fourth Plan target of 10 percent GDP
growth was surpassed by 1.2 percent, while 1971 and 1972 showed a
growth of over 14 percent in constant prices. Tables A.1, A.2, and
A.3 show the growth of the economy in the 1960-72 period and the
anticipated changes in the structure of the economy in the Fifth Plan.
GNP growth in the Fifth Plan is expected to reach nearly 26 percent
per annum compared to 11.2 percent in the Fourth Plan at constant
prices, reaching over $55 billion in 1977-78. Per capita income
rose from $384 in 1967 to $556 in 1972 and is expected to reach
$1,521 by the end of the Fifth Plan.

 The largest contributor to the growth of GDP has been the oil
sector as shown in Table 6.7.

TABLE 6.6

Summary of Iran's Balance of Payments
During the Fifth Plan
(billions of dollars)

1.	Current receipts		114.0
	(a) Receipts from oil sector	102.2	
	(b) Foreign exchange earnings from export of goods	4.9	
	(c) Foreign exchange earnings from export of services	4.9	
	(d) Foreign exchange earnings from investments abroad	2.0	
2.	Current payments		94.7
	(a) Sale of foreign exchange for import of goods	79.1	
	(b) Sale of foreign exchange for import of services	14.3	
	(c) Servicing of foreign loans	1.3	
3.	Current balance		19.3
4.	Receipts on capital account		4.7
	(a) Foreign loans and credits received by the government	2.2	
	(b) Other loans and foreign private investment	2.5	
5.	Payments on capital account		6.5
	(a) Repayment of principal of government loans and credits	6.0	
	(b) Repayment of private loans and transfer abroad of private capital	0.5	
6.	Capital balance		-1.8
7.	Net balance on current and capital account		17.5

Source: Fifth Development Plan of Iran (revised), Plan and Budget Organization (Tehran, 1975), p. 62.

TABLE 6.7

Sectoral Contribution to the Gross Domestic Product
at Constant Prices
(percent)

Sector	End of Third Plan (1967)	End of Fourth Plan (1972)	End of Fifth Plan (1977)
Agriculture	24.5	18.1	8.0
Oil	13.8	19.5	48.7
Industry and mining	21.3	22.3	16.1
Services	40.4	40.1	27.2
GDP at factor cost	100.0	100.0	100.0

Source: Fifth Development Plan of Iran (revised), Plan and
Budget Organization (Tehran, 1975), p. 37.

It can be seen that during the Fifth Plan nearly half of the gross
domestic product is expected to come from the oil sector compared
to 13.8 percent in 1967. The share of agriculture falls substantially
from 24.5 percent to 8 percent during this period, indicating a structural change in the economy. The role of the services sector must
not be underestimated: in the Third and Fourth Plan it contributed
40 percent to the GDP, and in the Fifth Plan it is still expected to be
far ahead of industry and agriculture.

Clearly the Iranian economy has been well served by the oil
receipts, although, because of the nature of the industry, these receipts have not required large investments in recent years. During
the Fourth Plan a total of $750 million was invested in the oil and
gas industry. Only $140 million or 19 percent of these investments
were allocated to the oil sector by the Plan Organization. The Iranian Oil Operating Companies (the Consortium) invested a total of
178 million pounds sterling, or approximately $445 million in the
southern oil operations, bringing the total oil investments in the
Fourth Plan up to $565 million. [5] During the Fifth Plan, however,
much larger investments are planned. Investment in the oil sector
is expected to reach $9,300 million--an increase of over 16 times
over the previous plan. Investment in the gas sector is also expected
to increase by four times to $2,500 million (see chapters 7 and 8).

DIRECT IMPACT OF THE OIL INDUSTRY

In the previous section we discussed the impact of the oil revenues on the Iranian economy; in this section we will consider the relationship between the industry itself and the various economic sectors through backward and forward linkages.

Forward Linkages

Forward linkages represent the flow of low cost fuel from the oil industry to the national economy. The cheap source of fuel was expected to provide an inducement for the economy to substitute oil for other energy resources and to create petroleum-based industries. This prediction has turned out to be true in the case of Iran, where in 1972 nearly 70 percent of the domestic energy requirements were supplied by the oil industry. If we add natural gas and LPG, the ratio rises to around 90 percent for most of 1960-72. The demand for petroleum products has risen nearly 17 times in 1950-74, while fuel oil and gas oil, the two most important industrial fuels, have shown very high rates of growth in this period.

The demand for petroleum products depends on various factors, such as the rate of growth of GNP and income per head, income elasticity of demand, prices, and the like. Since these subjects are discussed later (see Chapter 10), it is sufficient to note that there is a very close correlation between the demand for oil products and the rate of economic growth. There is what one may call a "cause and effect" relationship between the consumption of oil products and the rate of economic growth--that is, while economic prosperity is brought about by the growth of the oil industry, economic progress will in turn increase the demand for the products of the oil industry, which will lead to further growth of the oil sector.

The availability of easily accessible quantities of gas induced the government to build the 1,100-kilometer pipeline to the Soviet Union. This pipeline was completed in 1971 and serves two purposes: first, it is used to export gas to Russia, which in exchange paid for the Russian construction of the Esfahan steel mill, and second, it will provide natural gas for the main centers of consumption. The pipeline route has been drawn so as to serve the latter objective. Furthermore, the National Iranian Petrochemical Company (subsidiary of NIOC) has undertaken several projects for the construction of petrochemical plants, some in partnership with foreign companies. Some of these plants are completed and others are under construction. Although no important linkage in the petrochemical industry has yet materialized, it is fair to anticipate that in the next decades petrochemicals will become an important export activity. At the same time they will supply the domestic market with cheap fertilizers, plastics, and related products.

It is important at this stage to emphasize that in a typical dual economy one would expect the forward linkages to be weaker than the backward linkages. For instance, if the leading dynamic sector is producing final goods (such as automobiles), the backward linkages, or demand-induced influences, may become predominant. In extractive and mineral industries, the leading sector may provide its dominant influence through supply--that is, forward linkages. In the case of Iran, it is fair to say that forward linkages have in the post-nationalization period led to "some" integration between the national economy and the oil sector.

Backward Linkages

Backward linkages constitute the leading industry's demand for goods and services provided by the national economy. The leading sector's demand can be divided into (a) demand for fixed assets required for expansion of industry, and (b) demand for current resources to meet the routine requirements of industry.

Backward linkages, with regard to capital expenditure, have been particularly weak. The oil industry is a very capital intensive industry, and the degree of capital intensity is rising over time by the installation of sophisticated and automatic machinery in refineries and oil terminals. The national economy has thus far been unable to provide such capital equipment for the industry. There are, moreover, spillover effects from the import of this machinery from abroad insofar as the installation of this machinery would require the building of new plants (from domestic resources) and employment of domestic labor. But such spillovers are generally short-lived and really irrelevant to the issue of capital expenditures. The inability of the domestic economy to supply the capital goods required by the industry may not continue indefinitely, although the diversion of economic resources toward the production of such specialized facilities, in competition with large and experienced Western companies, may be irrational at this stage of economic development in Iran.

The only major part played by the domestic economy in supplying the needs of the oil industry for capital goods has been the construction of the Ahwaz pipe mill in 1968. This plant is producing pipelines for products and crude oil in various parts of the country. Although this plant has been recognized by the American Petroleum Institute and is authorized to use the standard mark of API, the pipes made in this plant may be more costly than the prices of pipes made in international markets.

The most important aspect of backward linkages is the current expenditure of the oil industry on the goods and services of the national economy. These expenditures include wage payments for labor, purchases of supplies, and industrial equipments from the indigenous sectors. According to a recent study during the 1961-68 period, the purchases of foreign goods by the Consortium rose by 25 times, while their purchase of domestically produced goods did not show any appreciable change during the same period. In contrast NIOC's domestic purchases rose in the same proportion as its foreign purchases. The purchases of NIOC are to a great extent peripheral to oil operations, involving the so-called "nonbasic operations" financed by the Consortium but carried out by NIOC. The "nonbasic operations" include housing (30 percent), medical services (about 20 percent), and administration (15 percent). The range of the non-basic activities are very diverse. They are mainly directed toward improving the living standards of the oil industry employees. In some cases they include swimming pools, clubs, holiday camps by the Caspian Sea, and school and road construction.

Employment

The oil industry's demand for labor is another backward linkage. Because of the capital intensive nature of the oil industry, the level of employment has actually fallen while the production has increased. Table 6.8 shows employment and productivity in the Iranian oil industry, 1958-71. The productivity in thousands of cubic meters per employee has risen from 0.50 to 7.03, an increase of about 14 times. Two other interesting points emerge from Table 6.8. First, despite the reduction in the labor force of the oil industry, the number of Iranian staff has nearly doubled in this period, indicating the expansion of NIOC and the substitution of Iranians for foreign staff. Second, the drastic reduction in the manual labor force during this period illustrates the efforts of the oil industry to raise productivity and introduce automation in the production and refining activities.

Table 6.9 shows the employment by the Consortium since 1958, while Table 6.10 shows the division of employment by various operators. It can be seen that employment by the Consortium has greatly diminished over this period from over 44,000 to nearly 18,000.

The reason for the decline in the Consortium's employment is not the gradual takeover of the operations by NIOC, but rather the increasing efficiency and automation of the operations. NIOC and the Consortium followed separate activities until July 1973, when the operation of the latter was taken over by the former. By March 1973 NIOC was the largest employer, followed by the Consortium. The

TABLE 6.8

Employment and Productivity in the Iranian Oil Industry
in Selected Years

| Year | Staff | | Manual Labor | Contractor[a] | Total | Production in Thousands of Cubic Meters | Productivity in Thousands of Cubic Meters/Employee |
	Iranian	Foreign					
1958	8,139	693	48,477	4,724	62,033	47,767	0.77
1961	10,188	847	39,638	1,619	52,292	68,581	1.30
1964	9,888	474	31,564	727	42,653	98,343	2.30
1967	11,659[b]	--	29,426	1,385	42,470	150,681	3.50
1970	12,547[b]	--	26,952	1,917	41,416	222,180	5.40
1972	12,831	497	24,931	2,766	41,812	294,100	7.03

[a]Excluding foreign employees of the contractors.

[b]Including foreign staff.

Source: For employment figures up to 1967, M. Nezam-Mafi, "Role of Oil in the Iranian Economy," booklet published by NIOC Public Relations Office, 1967. For other data, NIOC's Statistical and Information Office of the Affiliated Companies.

145

TABLE 6.9

Employment in the Iranian Consortium in Selected Years

Year	Field Operations		Abadan Operations		Tehran Office		Total[a]
	Iranian	Foreign	Iranian	Foreign	Iranian	Foreign	
1958	18,977	223	24,661	258	181	110	44,410
1961	14,962	263	22,634	346	428	186	38,918
1964	11,744	181	18,016	112	554	101	30,708
1967	9,983	143	14,979	88	628	97	25,918
1970	7,114	163	10,818	45	766	118	19,024
1972	6,952	b	9,923	b	952	b	17,827

[a]Includes the personnel in nonbasic operations. Nonbasic personnel amounted to around 40 percent of the Consortium labor force.

[b]No Iranian/foreign division was made in 1972.

Source: Iranian Oil Operating Companies, Annual Reports, 1963, 1967, and 1972.

TABLE 6.10

Number of Employees by Various Operators at the End of March 1973

Company	Iranian Employees						Expatriates		Total Work Force
	Posted		Contractors		Temporary		Co.		
	Staff	DRE's*	Staff	DRE's*	Staff	DRE's*	Staff	Contractors	
National Iranian Oil Company	5,138	9,414	639	113	11	179	31	338	15,863
Nonbasic operations	2,132	4,564	75	23	--	--	17	--	6,811
National Petrochemical Company and subsidiaries	752	526	692	1,381	90	266	2	79	3,788
National Iranian Gas Company	1,088	1,342	105	--	27	--	29	--	2,591
Ahvaz Pipe Rolling Mill	71	197	6	--	1	--	2	--	277
National Iranian Tanker Company	17	--	13	--	--	--	--	--	30
Iranian Oil Operating Companies	3,226	7,544	16	--	--	--	305	283	11,374
SIRIP	152	118	6	27	2	--	12	--	317
IPAC	215	50	3	1	--	--	16	61	346
IMINOCO	209	96	7	13	--	--	23	65	313
IROPCO	1	--	--	--	--	--	1	--	2
LAPCO	142	124	26	6	--	--	23	7	328
PEGUPCO	1	--	1	2	--	--	1	--	5
SOFIRAN	41	9	6	2	--	--	17	--	75
CONIRAN	15	4	--	1	2	--	7	18	47
BUSHCO	14	1	3	2	3	2	7	54	86
HOPECO	6	--	--	--	--	--	--	--	6
INPECO	36	6	--	6	--	--	18	--	66
Total	13,156	23,995	1,598	1,577	136	447	511	905	42,325

*Daily rate employees.

Source: Iran Oil Journal, June 1973.

number of people employed by the joint ventures or service contracts
was very small. The largest employer among the joint ventures was
IPAC, with 346 employees or less than one-tenth of 1 percent of the
total. The companies which have not yet discovered oil in commercial quantities have a labor force ranging from 2 (IROPCO) to 75
(SOFIRAN).

The total active labor force in Iran was 8.3 million in 1973-74,
of whom 7.6 million were employed. In the same year the number
of industrial labor force amounted to around 2 million.[6] This means
that just over one-half of 1 percent of the employed labor force was
working for the oil industry, and the share of the oil industry out of
the total industrial labor force was 2.1 percent.

One final point which may be mentioned is that, although no
accurate data is available on the outflow of trained manpower from
the oil industry to the domestic economy, the author's inquiries point
to some transfer of technical and managerial skills from the oil industry to the national economy. The outflow of qualified manpower
from the oil industry has in many cases been due to relatively low
salaries of Iranian graduates in the oil sector compared with those
of other government departments, universities, or private enterprise.

The overall impact of the oil industry in bringing about an integration with the domestic economy has been less important than the
fiscal effect of the oil revenues. While forward linkages have become
increasingly important by supplying cheap energy to the economy, the
backward linkages have had no substantial effect.

The interrelationship between the oil industry and the economy
may be observed through the input-output relationships. The only
available input-output table was published in the mid-1960s and because of lack of information is not very illuminating. However, on
the basis of the table it can be seen that to produce 100 rials of output
the oil industry buys 7.3 rials of goods from other sectors and in
turn supplies these sectors with 8.4 rials of petroleum products.[7]

SUMMARY AND CONCLUSION

In this chapter we have looked at the direct and indirect effects
of the oil industry on the Iranian economy. It was shown that the
fiscal impacts of the oil industry in terms of revenues received by
the government had the most important effect on the Iranian economy
through the various development plans. The direct impact of the oil
industry, via forward and backward linkages was small--with the
exception of the flow of cheap fuel to the domestic economy.

The conclusions reached in Chapter 1 for 1901-51 were that
neither direct nor indirect influences had any impact on the Iranian

economy. The payments to the government were often misused by
the rulers, while the size of the payments were relatively small compared to the other sources of government revenue. In 1937 only 13
percent of the government revenues came from the oil receipts. This
ratio was 11.5 percent in 1949. There was no comprehensive development planning except for certain "impact" projects, such as the
construction of the trans-Iranian railway system. In 1949 only 14
percent of the total government expenditure was spent on development
projects. Unlike the 1901-51 period, the 1951-74 period saw a complete change in the Iranian oil industry.

Development planning started on a moderate scale in 1949 and
reached comprehensive and sophisticated levels in the Fourth Development Plan. Production and revenues rose greatly. The oil revenues reached a level of $18.0 billion in 1974 compared to $34.4 million in 1954, an increase of over 500 times. These large amounts of
revenue did not go to make any private fortunes, nor were they invested in foreign banks for speculative or other purposes. The
Iranian government utilized these revenues to the full. Indeed, in
many years Iran had a balance-of-payment deficit and had to resort
to large-scale borrowing, because the oil revenues were not sufficient
for the ambitious development plans of the country.

The division of the oil revenues by the government, between the
Treasury General and the Plan Organization, shows the attitude of
the government toward development planning. In the 1950s, more
than half of the oil receipts went to the ordinary budget, while in the
late 1960s over three-quarters of the oil receipts were allocated to
the development budget. The results, therefore, point to a fast rate
of economic growth: 9 percent in 1963-70 and over 14 percent in the
1971-72 period, at constant prices. After 1972 the pattern of growth
was disrupted by increasing oil revenues and growth rates of 33.8
percent in 1973 and up to 40 percent in 1974 were reported.

The direct impact of the oil industry on the economy was small.
While the forward linkages became effective over this period, the
backward linkages failed to materialize. The reason for this lack of
appreciable interrelationship between the oil industry and the economy
is not hard to find. The capital intensive nature of the oil industry
and the inability of the domestic economy to supply the sophisticated
machinery required by the oil sector have been the major factors.
The Consortium itself did not try to create by-product industries
which supply the requirements of the oil industry and consistently
relied on imported goods.

The pattern of development of the Iranian economy, as in many
other developing countries, does not match any particular development theory. It is but one more "special case" and it is wrong to
generalize from the pattern, even for other Middle East oil economies.

But why did the oil industry not have a greater impact on the prenationalization Iranian economy? And was the nationalization itself a
significant factor in paving the way for the oil industry to influence
the economy in such a way? One cannot claim that nationalization on
its own changed the situation. One may even argue that the nationalization actually hampered development planning by cutting off the oil
revenues to the First Plan. Clearly there were other factors at work,
both economic and noneconomic. A massive inflow of cash cannot by
itself transform an underdeveloped economy into a developed one.
In the postnationalization period large quantities of foreign exchange
became available. These resources were used wisely and under pre-
cise and comprehensive economic planning. Large investments in the
establishment of an infrastructure, capable of absorbing such vast
quantities of cash, were undertaken. Transport, communications,
and health and education were among the top priorities of the government.

In the first three plans a substantial portion--two-thirds of the
total in the First Plan, and over two-fifths in the Second and Third
Plans--was devoted to transport, communications, health, and education. In the Fourth Plan, however, the proportion devoted to the
economic and social infrastructure fell to 25 percent. An important
noneconomic factor has been the enlightened political leadership which
Iran has received from the Shah. The Consortium negotiated with
him whenever there was a dispute and foreign capital found a safe and
secure place for investment in Iran. The political stability enjoyed
by Iran provided a great incentive for entrepreneurs to set up industries in the country, and there was a general mood of confidence in
business circles. The role of the government in channeling the funds
into appropriate development expenditures must not be underestimated.
The pattern of economic development in Iran, unlike the Schumputerian
theory which draws its motivating power from the sphere of entrepre-
neurial decisions, is based not so much on innovation and management
initiatives, but mainly on the ability of the government to utilize the
borrowed technology and the available funds in a correct and productive
manner.[8]

Iran has obviously been able to achieve a high rate of growth by
properly utilizing its oil revenues. But this progress did not come
overnight. It took 20 years for Iran to build an infrastructure capable
of absorbing the oil revenues, and it took as long for the government
to improve its methods of employing these revenues. The Iranian
economy has developed a pattern peculiar to itself. Perhaps the future
development theorist will present a general theory which will incor-
porate the pattern of economic development in Iran.

NOTES

1. For a good analysis, see J. Amuzegar and A. Fekrat, Iran: Economic Development Under Dualistic Conditions (Chicago: University of Chicago Press, 1971); Robert E. Looney, The Economic Development of Iran (New York: Praeger Publishers, 1974); J. Bharier, Economic Development in Iran, 1900-70 (London: Oxford University Press, 1971); and M. Fardi, "A Macro-economic Model of Iran," unpublished Ph.D. dissertation, University of Illinois at Urbana-Champaign, 1972.

2. Bank Markazi Iran, Annual Report and Balance Sheet, 1970-73.

3. Ibid.

4. For a thorough documentation, see Amuzegar and Fekrat, op. cit., chap. 3.

5. Iranian Oil Operating Companies, Annual Reports, 1971 and 1972; Bank Markazi Iran, op. cit., Statistical Appendix, 1970-73.

6. Bank Markazi Iran, op. cit., 1973.

7. Fardi, op. cit.

8. Amuzegar and Fekrat, op. cit., p. 70.

7

THE PETRODOLLAR
PROBLEM

The quantum increase in oil prices in the last quarter of 1973 created a completely new situation for the Iranian political and economic strategy--a situation not anticipated by experts and one for which both the producing and consuming nations were unprepared. This chapter proposes to provide an introductory note to the petrodollar problem on a global scale and to examine Iran's alternatives for domestic and external expenditure of such funds.

MAGNITUDE OF THE PROBLEM

The OPEC members' receipts of oil revenues in 1974 amounted to $90 billion--four times their receipts in 1973 as shown below. It can be seen that small Arab producers are all in the "rich oil countries" group. With only 4 percent of the total population they control 45 percent of the OPEC revenue. These are the countries with the lowest level of absorptive capacity and the highest magnitude of surplus petrodollars. The nonoil exports which can be a rough indicator of the stage of economic development in these countries, shows that the five richest countries' nonoil exports were 21 percent less than Iran's and about one-third of Indonesia's nonoil exports.

In 1973 OPEC imports of goods and services amounted to $28 billion, up a solid 50 percent from 1972. In 1974 OPEC imports were estimated at $45-50 billion--a two-third increase over 1973-- in part as a result of higher world prices for manufactures and food. For 1975 these countries are expected to import $65-70 billion worth of goods and services.[1] These expenditures would create surpluses of about $45-50 billion in 1974 alone. Various estimates have been made as to the extent of these surpluses in future years. A World

Bank study estimates oil-producer country surplus to reach $650
billion by 1980, while OECD (Organization for Economic Cooperation and Development) forecasts accumulated holdings of $300 billion
by the same year.[2]
    Indeed, as we shall see later, the fall in demand and production
of oil in 1974 and the first quarter of 1975 indicates that even OECD
estimates are exaggerated, particularly in view of the freeze on oil
prices at a time of high inflation in the West.
    But what do all these figures mean in a global context? In 1974
two main schools of thought developed around the petrodollar problem. A number of economists like Walter Levy foresaw recession
and chaos on a global scale; others like Gerald Pollack and Thomas
Stauffer placed more faith in the "invisible hand" of the free market
--which could perhaps deal with the situation with a little intervention.[3] These arguments and counterarguments resulted in a number
of myths and misunderstandings surrounding the petrodollar question.[4] Some of these myths were the consequence of ignorance or
hasty doomsday predictions, while others were deliberate political
ploys employed by the supporters of Israel in the West, and particularly in the United States, who saw the increasing Arab wealth as a
direct threat against Israel, as well as those who believed in a more
restrictive trade policy.
    Let us summarize the main problems created by such sur-
pluses, examine their validity and offer remedies in some cases.

Possibility of a Fund Cutoff

    Some people argued that a number of OPEC countries might cut
off the flow of funds to the West in the same way as they cut off their
oil during the embargo period. This argument is false. As soon as
oil is produced in OPEC countries, the funds are automatically recycled. This is so because the payment for oil is nothing but a credit
transfer from a number of banks to others (or even in the same bank).
Thus the money cannot be physically shipped out and hidden under the
desert sands! The oil producers will draw on these funds which are
kept in the industrial countries, for imports, defense, investments
abroad in physical and financial assets, aid and loans to underdeveloped and industrial nations. This means that the aggregate quantity
of money in the industrial nations will not substantially change. Thus
the problem is not whether the petrofunds are going to be recycled or
not, but rather the way such recycling is going to take place.

TABLE 7.1

Estimated OPEC Oil Revenues and GNP, 1973-74

Country	Population (millions) 1973[a]	Oil Revenues (millions of U.S. dollars) 1973	Oil Revenues (millions of U.S. dollars) 1974	Nonoil Exports 1973	GNP per Capita[b] (U.S. dollars) 1973	GNP per Capita[b] (U.S. dollars) 1974
Rich oil countries	11.2	10,600	40,300	683		
Saudi Arabia	7.8	5,100	20,000	577	980	2,900
Kuwait	1.0	1,900	7,000		4,100	8,500
Qatar	0.2	400	1,600		3,300	>10,000
Abu Dhabi[c]	0.1	900	4,100		9,000	>10,000
Libya	2.1	2,300	7,600	106	3,000	5,800
Middle-income oil countries	75.2	9,500	39,700	2,388		
Iran	31.2	4,100	17,400	864	520	940
Venezuela	11.2	2,800	10,600	1,232	1,150	1,850
Iraq	10.4	1,500	6,800	22	430	930
Algeria	15.4	900	3,700	270	350	530
Ecuador	6.5	100	800	n.a.	320	420
Gabon	0.5	100	400	n.a.	900	1,540
Low-income oil countries	183.4	2,900	10,000	2,210		
Indonesia	124.0	900	3,000	1,923	80	100
Nigeria	59.4	2,000	7,000	287	150	230
Total	269.8	23,000	90,000	5,281		

[a]Population figures extrapolated from 1971 data as shown in World Bank Atlas.
[b]GNP figures were estimated on the basis of both oil revenues and nonoil productive activities.
[c]Member of the United Arab Emirates.

Source: IMF Survey, February 3, 1975, p. 38.

Inflation Impact

The increase in oil prices which resulted in accumulation of
such funds has been frequently blamed as the main cause of the high
rate of Western inflation. This argument is also not credible. The
inflationary pressures had already developed in Western countries
by the late 1960s and early 1970s as a result of the Vietnam War in
the case of the United States and wage-push inflation in Europe. Two
respectable studies in 1974 show the inflationary effect of the oil
prices (Table 7.2). Although the two estimates are not identical,
they roughly show the order of magnitude of the problem--only a few
percentage points of inflation in the West is attributable to the increased oil prices.

TABLE 7.2

Inflationary Impact of Oil Prices, 1974

Country	Overall Rate of Inflation, Percent	Inflationary Effect of Oil Prices, Percent	
		(1)	(2)
United States	12.0	--	0.8
Japan	23.0	--	1.8
France	14.9	2.5	1.7
West Germany	5.9	1.5	1.7
Italy	25.2	3.6	2.3
United Kingdom	18.3	2.7	2.4
Netherlands	11.0	1.5	2.5
Benelux	15.7	4.2	--
Denmark	16.5	1.9	--
Ireland	20.0	1.7	--
Total OECD	--	--	1.8

Note: Based on OECD data.

Source: (1) Based on the report by European Economic Com-
munity--Economic and Social Committee, 1974; (2) Thomas R.
Stauffer, "Oil Money and World Money: Conflict or Confluence?"
Science 184, no. 19 (April 1974).

Financial Destabilization

It is argued that surplus funds when placed in short-term deposits may lead to financial destabilization (through bank default or weakening of the currencies) if and when they are withdrawn. Furthermore, individual banks may find it imprudent to handle short-term high interest deposits and lend to long-term borrowers, because of the potential instability of the deposit base: when a few oil producers control a substantial and growing share of the market's deposits.

In the first place there is no evidence to suggest that OPEC members will for political or irrational purposes move their deposits around. The only recent experience is that of Libya in 1971 when Colonel Qaddafi decided to move the Libyan holding of 200 million pounds sterling from Britain to Switzerland. This was meant as a retaliatory act against what he charged was "Anglo-Iranian collaboration" on the return of Abu Musa and smaller and larger Tumb Islands in the Persian Gulf to Iran. This action is alleged to have cost Libya up to 5 percent of her holdings due to currency fluctuations and brokerage fees. It is therefore clear to all holders of petrofunds that any such movements on a large scale is detrimental to their holdings. At the same time, short-term deposits are gradually replaced by longer-term deposits, while the remaining short-term deposits are being automatically renewed on maturity--making them long-term in nature.

Moreover, it is true that individual banks may be subject to this kind of danger, but this really is not an OPEC problem. It is a shortcoming of the international financial system itself, which has been placed under spotlights because of the flow of petrodollars. Some system of coordination and cooperation among the financial institutions of the OECD countries is needed to limit the possibility of such potential dangers. The Eurocurrency markets have, however, demonstrated their flexibility and sophistication for handling such unprecedented strains on the international payments system. Two particular methods were often employed to deal with this problem: loan syndication whereby a group of banks provide the loans to spread default risk, and rollover clauses which provide a balance between borrowing and lending terms to minimize withdrawal risks. According to William Simon, U.S. Secretary of the Treasury,

> A few individual institutions in the United States
> and elsewhere did in fact experience difficulties.
> But their troubles arose mainly from internal
> management problems which came to the fore in
> an environment of inflation, restrictive money

policies and generally rising interest rates, or from
their own failure to exercise proper supervision as
they rapidly expanded their foreign exchange trading.
The difficulties they experienced were not the result
of massive inflow of OPEC monies. [5]

Simon went on to say that he believed that the petrodollar problem is
"manageable. "

Uneven Reallocation of Surpluses

One problem often discussed is that petrodollars are not recycled to those countries that need them most, since these funds were
expected to flow to those countries which were in a relatively stronger
position in the first place. The capital markets that received these
funds were not prepared to lend to the countries which already had
substantial payments deficit for fear of default. An often-quoted
example was Italy with its declining creditworthiness in the Eurocurrency markets, as well as many underdeveloped countries. To
get the problem in a proper perspective, let us see what happened to
the funds in 1974. According to the U.S. Treasury, which believes
the surplus to have been $60 billion:

- $21 billion, or about 35 percent of the surplus, went into the
  Eurocurrency market, basically in the form of bank deposits.
- $11 billion, or 18.5 percent of the total, flowed directly into
  the United States. It is estimated that roughly $6 billion went
  into short-term and long-term U.S. government securities,
  $4 billion were placed in bank deposits, negotiable certifi-
  cates of deposit, and other money market paper. Less than
  $1 billion was invested in property and equities in the United
  States.
- $7.5 billion, or 12.5 percent, is believed to have been in-
  vested in pound sterling denominated assets in the United
  Kingdom, some of it in U.K. government securities, some
  in bank deposits, some in other money-market instruments,
  and some in properties and equities. This of course is quite
  apart from the large Eurocurrency deposits there.
- $5.5 billion, or about 9 percent, may have been accounted
  for by direct lending by OPEC countries to official and quasi-
  official institutions in the developed countries other than the
  United States and the United Kingdom.
- $3.5 billion, or 6 percent of the total, represented OPEC
  investment in the official international financing institutions
  such as the World Bank and the IMF.

- $2.5 billion, or 4 percent, has flowed from the OPEC countries to other developing countries, both bilaterally or through various OPEC lending institutions.
- $9 billion, or 15 percent, cannot be accurately identified, but this residual would cover funds directed to investment management accounts as well as private sector loans and purchases of corporate securities in Europe and Japan.[6]

The above figures suggest that the private banking system of the industrial countries probably accepted half the total surplus in 1974, with the major role played by the Eurocurrency banks. In the words of Mr. Simon, "The relatively balanced pattern of OPEC investment explains in part why the massive shift in financial assets did not lead to the financial crises that some envisioned."[7]

The year 1974 proved to be a year when all the doomsday scenarios fell apart, thanks to the rational policy of OPEC investment, flexible exchange rates, and the sophistication of the Eurocurrency market. This is not to say that there was no danger of a financial crisis. The industrial countries have long been aware of the need for a reform in the international monetary system, but the extent of the danger came to light in view of the potential petrodollar problem. One can only hope that coordinating machineries (such as the Bank of International Settlement) will be used in the future to protect the countries in severe temporary payments trouble, through the so-called "swap arrangements." This will help reduce the danger of "beggar-my-neighbor" policies which were fortunately avoided in 1974. If all countries decide to reduce their payments deficit through import restriction or export subsidies a trade war with far-reaching economic and political implications will become inevitable. In 1974 no other major industrial power risked such a policy, except for Italy, which undertook restrictive trade policies (and was forgiven because of her difficult situation). In early 1975 some progress was made toward official international cooperation to strengthen multilateral facilities. How efficient these mechanisms will be only time can tell us.

The agreements included a $25 billion solidarity fund along the U.S. proposed "safety net" proposal to provide supplementary financing, if the need arises, to participating OECD countries which follow cooperative energy and economic policies. Funds will be provided by participants and through borrowing of OPEC funds indirectly so as to reduce OPEC's exercise of political influence. IMF Oil Facility was $3.6 billion in 1974; in 1975 it was agreed to increase this to over $6 billion. It was also decided to use contributions from oil producers and industrial countries in order to subsidize interest costs of the oil facility for the very poorest countries. The United

States opposed the expansion of oil facilities to larger volumes, because she viewed this facility as a forum on which OPEC could exercise political or economic control, as a result of its direct contributions. Other agreements included a one-third increase in IMF quotas to give the organization nearly $50 billion to help its 126 members. The oil exporters' collective quota was to double in order to provide for greater participation and a greater voice in the activities of the Fund.

Impact on Developing Nations

Much was said in 1974 about the impact of the oil price increase on the developing countries and how OPEC should spend its petrodollars helping the so-called Fourth World nations. The newly found concern on the part of the industrial nations for the poor countries is amusing. During the whole of the first Development Decade (1960-70) and part of the second (1970-80), the poor countries struggled at various UNCTAD (United Nations Conference on Trade and Development) meetings for preferential tariff privileges and reduction in nontariff barriers without much success in the face of opposition from Western countries, and particularly the United States.[8] The industrial countries that pledged 1 percent of their GNP for aid to poor countries never even came close to the target.[9] This does not mean that the problem was not serious; indeed the oil price increase placed a very heavy burden on the balance of payments of the developing countries. Not only had the new situation reduced the ability of the industrial countries to give aid, but also the poor countries faced great difficulties in raising funds in the capital markets.

According to various reports from the World Bank, the developing countries required $4-5 billion in 1974 to finance their oil deficits and around $9 billion in 1975. Such financing may originate from four major sources: increase in exports, foreign borrowing on capital markets, multilateral and bilateral loans, and aid. The possibility of export expansion in short-term, particularly when the development of these countries is set back by the escalating oil prices, is relatively limited. Borrowing in capital markets has also become more difficult for these countries, because financing of oil-related deficits is relatively unattractive to the lenders. Many developing countries, particularly those with high debt-service burdens and without rich mineral endowments are unable to borrow, and more are likely to be excluded as their debts grow. Thus the financing needs of these countries will increasingly have to be handled outside the market. Table 7.3 provides a summary.

TABLE 7.3

Publicized Eurocurrency Credits, 1972-74
(medium- and long-term loans in billions
of dollars or equivalent)

Borrower	1972	1973	January-June 1974
Developed countries	4.32	11.12	13.07
Developing countries*	3.80	9.12	6.01
Nonmembers of IBRD	0.40	1.78	0.66
Total	8.52	22.02	19.74

*Includes countries not members of the World Bank, oil exporters, other international organizations, and borrowers unallocated by country.

Source: International Financial Division, International Bank for Reconstruction and Development, August 1974.

As can be seen from the table, there was a marked shift in the composition of the Eurocurrency loans during the first half of 1974. Although the total increase in the loans was substantial (due to inflow of petrofunds), the distribution of the loans was detrimental to the developing countries. The share of the developing countries out of the total credits declined by one-third between 1972 and mid-1974-- from 45 to 30 percent. Even these figures do not clearly reflect the extent of the problem: in the first half of 1974 the credits to three European borrowers--United Kingdom ($4.8 billion), France ($2.9 billion), and Italy ($2.2 billion)--amounted to half the total credits. For underdeveloped countries the same pattern emerges: countries such as Spain ($0.85 billion), Mexico ($0.98 billion), Brazil ($0.62 billion), and the Philippines ($0.78 billion) which were already in a stronger situation received 54 percent of the credits available to the developing countries. The share of African countries, for example, declined from 10 percent in 1973 to 3 percent in the first half of 1974.[10]

OPEC's major aid contribution to developing countries has been through multilateral and bilateral aid, and this type of aid is expected to be dominant in the future. Most OPEC members are newcomers in the aid league, except for Kuwait which has been an aid donor for more than a decade. The flow of official economic aid from OPEC countries was between $380 million and $530 million annually during 1970-73. Cumulative disbursements since the end of the 1967 Middle

East war have been $3 billion (excluding $600 million IBRD bonds),
essentially from Kuwait, Saudi Arabia, and Libya. Until 1973 about
$2.5 billion or 83 percent of the aid disbursements were directed to
Egypt, Syria, and Jordan. Until 1973 the above three countries were
the only producers providing aid on a significant scale. Their total
net disbursement, excluding military assistance, amounted to $408
million in 1972 and $491 million in 1973, corresponding to 3.6 per-
cent and 2.7 percent respectively of their gross national product.
In general, OPEC contribution to multilateral organization before
1974 has been negligible or nonexistent.

It is important at this stage to make a distinction between committed and disbursed aid. Unfortunately, most of the information
about aid comes from press reports which do not make such a distinction. (Table 7.4 provides a breakdown of OPEC aid.) The difference is very important because committed aid may never actually
be paid. A letter of intent signifies committed aid, but a host of
other conditions at various stages in time must be met before the
committed aid becomes actual. For instance, an agreement may
stipulate that aid will be granted if there is no significant change in
the domestic and international financial position of the donor. This
condition would give the donor an opportunity for withdrawal.

Another important characteristic of aid is its so-called "grant
element," as well as the sources to which it is tied. Aid is defined
as outright grant (nonreturnable), loans, and sometimes foreign
private and public investment. Most of this aid is tied to political
and economic factors--where for instance, the host country has to
spend part or all of its aid in the donor country. Moreover, loans
have three main features: grace period, interest rate, and repayment period. By changing any of these three variables the grant
element (what the aid is really worth to the host country as grant)
will completely change. [11] In this way OPEC aid is significantly different to Western aid in nature. Unlike U.S. or European aid which
is almost always tied to purchases in these countries, reciprocity in
trade, or to particular projects (Sweden and West Germany are
sometimes exceptions), OPEC aid is more flexible. This aid in no
case requires purchases in the donor countries, but occasionally it
is related to specific projects. The so-called project-tied aid from
OPEC is again different to similar aid from the industrial countries.
Kuwait Fund for Arab Economic Development operates like the World
Bank. It requires feasibility studies for specific projects which the
Fund believes to be most beneficial to the receiving country. Others,
such as Saudi Arabia, United Arab Emirate, and Libya also give
project aid, although they are not as sophisticated as the Kuwaitis
in their project evaluation. More developed OPEC countries, particularly Iran and Iraq, are interested in projects which may eventually

TABLE 7.4

Committed and Disbursed OPEC Aid, 1974*

(millions of dollars)

Donor Country	Commitments					Estimated Disbursements				
				As Percent of					As Percent of	
	Bilateral	Multi-lateral	Total	Oil Revenue	GNP	Bilateral	Multi-lateral	Total	Oil Revenue	GNP
Iran	2,802	173	2,975	17.1	10.1	600	2	602	3.4	2.0
Kuwait	957	384	1,341	19.1	15.8	460	70	530	7.6	6.2
Libya	178	241	419	5.5	3.4	45	60	105	1.4	0.9
Saudi Arabia	2,568	453	3,021	15.1	13.4	650	115	765	3.8	3.4
Venezuela	20	726	746	7.0	3.6	20	165	185	1.7	0.9
Algeria	31	108	139	3.8	1.7	15	45	60	1.1	0.7
Iraq	222	58	280	4.1	2.9	80	35	115	1.7	1.2
Nigeria	1	16	17	0.2	1	15	15	16	0.2	0.1
Qatar	97	60	157	9.8	7.4	40	20	60	3.7	2.8
United Arab Emirate	296	182	478	11.6	10.0	120	45	165	4.0	3.5
Total	7,172	2,401	9,273	12.1	8.2	2,031	572	2,603	3.3	2.2

*Provisional data for 1974. Multilateral aid excludes contributions to IMF and IBRD.

Source: Based on data from World Bank.

produce exports for their own markets, either as raw materials or
as intermediate goods.

In general, OPEC loans have a much greater grant element
than Western aid.   Grace periods and repayment periods are
longer, while interest rates are far below the free markets rates
(for example, Iran's loans to India at 2.5 percent interest). There
is also a great deal of grant aid given by the OPEC countries, which
is highly unusual in the Western context. Thus the comparison of the
absolute magnitude of OPEC and OECD aid is misleading, since the
two types of payments are different in nature.

As can be seen from Table 7.4, the total committed aid by
OPEC in 1974 amounted to over $9.2 billion or 8.2 percent of their
combined GNP. Saudi Arabia, Iran, and Kuwait were the largest
three donors both in terms of magnitude and percentage of GNP. The
disbursed aid was $2.6 billion or 28 percent of the commitments;
Kuwait actually disbursed 6.2 percent of her GNP, Saudi Arabia 2.4
percent, and Iran 2 percent. Among the more developed OPEC
countries Iran was by far the largest donor.

Multilateral OPEC aid was provided through two main channels:
the international organizations long established for aid and loans such
as IMF and IBRD; and a large number of smaller development funds
which surfaced in recent years. During January–September 1974,
OPEC countries made available $3.1 billion to the IMF oil facility
and $1 billion to the World Bank (these payments are not shown in
Table 7.4). Moreover, they committed the following sums: Islamic
Development Bank ($900 million); the Special Fund of the Inter-
American Development Bank--IDB ($500 million); the Arab Bank for
Africa ($250 million); the Special Arab Fund for Africa ($200 million);
the UN Emergency Fund ($150 million); the OPEC Development Fund
($150 million); and several other institutions ($250 million).[12] Out
of this $2.4 billion committed multilateral aid only $572 million was
disbursed in 1974.

Bilateral OPEC aid assumed a great deal of importance and is
expected to become more important in the next few years. The recipient countries included many Arab as well as non-Arab developing
nations, particularly those listed by the United Nations as MSA (most
seriously affected) countries. As Table 7.5 shows, the largest recipients of actual and committed aid were Egypt and Syria. African
non-Arab countries received less than they had anticipated as the
price of breaking off their diplomatic relations with Israel, although
their share of the actual payments was larger than their share in
committed aid.

It may be useful at this stage to make a brief comparison between the aid from developed countries of OECD and OPEC aid. In
1972 Official Development Aid (ODA) of OECD amounted to $8.6 billion,

TABLE 7.5

Recipients of Committed and Disbursed
OPEC Bilateral Aid, 1974[a]
(millions of dollars)

Recipients	Commitments		Estimated Disbursements	
	Millions of Dollars	Percent of Total	Millions of Dollars	Percent of Total
Egypt	3,121	43.5	765	37.1
Syria	1,003	14.0	325	16.0
Jordan	185	2.6	140	6.9
Mauritania[b]	153	2.1	25	1.2
Sudan[b]	107	1.5	70	3.4
Somalia[b]	82	1.1	25	1.2
Morocco	80	1.1	25	1.2
Tunisia	54	0.8	15	0.7
Bahrain	21	0.3	10	0.5
Yemen (A.R.)[b]	19	0.3	15	0.7
Yemen (P.P.R.)	12	0.2	10	0.5
Other Arab countries	8	0.1	8	0.4
Madagascar	114	1.6	--	--
Guinea[b]	16	0.2	5	0.2
Uganda	12	0.2	2	0.1
Senegal[b]	11	0.1	5	0.2
Other African countries[b]	79	1.1	46	2.3
Pakistan[b]	957	13.3	355	17.5
India[b]	945	13.2	75	3.7
Sri Lanka[b]	86	1.2	35	1.7
Bangladesh[b]	82	1.1	50	2.5
Guyana	15	0.2	15	0.7
Honduras	5	0.1	5	0.2
Malta	5	0.1	5	0.2
Total	7,172		2,031	

[a]Provisional Data for 1974.
[b]MSA (most seriously affected) countries.

Note: 37.4 percent of committed funds are going to MSA countries; 36.2 percent of disbursed funds are going to MSA countries.

Source: Based on data from the World Bank.

22 percent of which went through multilateral organizations. Other
aid including private investment brought the total up to $19.6 billion.
ODA is, however, the most important component from our point of
view. Since it is governmental aid it is in nature comparable to
OPEC aid. The ODA target of 1 percent of GNP promised by the
members of the Development Assistance Committee (DAC) is unlikely
to be reached by the previously agreed date of 1975. Table 7.6 shows
that on the whole the member countries contributed one-third of 1
percent of their GNP to developing countries in 1972. The grant element of their loans was 57 percent of the value of loans and the grant
element of the total ODA was 85 percent of its face value. This
means that the real contribution of these countries as a percentage
of their GNPs is lower than the figures shown in the table.

TABLE 7.6

Official Development Aid by the Development
Assistance Committee of OECD, 1972

| Country | ODA as Percent of GNP | Grant Element, Percent | |
		Loans	Total CDA
Australia	0.61	--	100
Austria	0.09	38	81
Belgium	0.55	63	96
Canada	0.47	90	97
Denmark	0.45	76	94
France	0.67	32	86
Germany	0.31	60	81
Italy	0.08	20	59
Japan	0.21	42	61
Netherlands	0.67	64	85
Norway	0.41	59	100
Portugal	1.91	32	51
Sweden	0.48	80	95
Switzerland	0.22	89	96
United Kingdom	0.40	64	86
United States	0.29	65	87
Total DAC	0.34	57	84

Source: Development Cooperation, OECD, 1973, pp. 44-49.

The OPEC countries have entered the big aid league in the span of one to two years. They have given aid generously to the poor without expecting a substantial feedback from these countries and without imperialist intentions. The Arab countries have given aid to those sympathetic toward their struggle against Israel, while Iran and other non-Arab producers have directed their aid to the expansion of their political influence as well as humanitarian ground. Such linkage between politics and aid is totally acceptable within the political framework of the recent times. Indeed, such aid has been more substantial than the Western aid both on economic grounds (percentage GNP and grant element) and on humanitarian grounds. To say that these countries should give all they cannot internally absorb in aid is hypocritical. Oil is an exhaustible resource which will run out sooner or later, and it is good business judgment to invest parts of it abroad for future returns and to expand domestic absorptive capacity. After all, neither the United States nor the major European countries gave as aid all their surpluses. They formed multinational corporations to go abroad in search of profitable ventures, while they expanded the comforts of their life in their home countries.

Financial versus Real Transfers

Surprisingly little distinction is made in the literature between financial and real transfer of wealth. A financial transfer is simply a credit transfer where no physical goods are exchanged. Physical transfer of resources from the consuming nations to producing countries is limited to the latter's import capacity and the producers' aid with which physical goods are bought. The financial assets are interest-bearing and add to the burden of the consuming countries, but the time lag involved for producers to expand their absorptive capacity will give the consumers an opportunity to expand their productive capacity, possibly take a gradual small cut in the living standards, and develop alternative sources of energy. If the producers could have immediately absorbed all their surplus (or give it all in aid), there would have been a full transfer of physical resources out of the industrial countries equivalent to the total OPEC receipts. This would have had serious economic implications in the industrial countries.

The economic recession of 1974-75 in the industrial countries has added to the confusion. Because of the fall in the general level of demand, it was assumed that if OPEC countries could import more and give more aid, the level of demand would rise and the economy would emerge from recession. Such assumptions are based on a misunderstanding of the needs of the developing countries. If there were such a capacity to import, it would not have been directed toward the U.S. auto industry, for example, but rather toward food, chemicals (naphtha), steel, cement, paper, and other capital goods which are already in short supply in the Western economies. Thus the lag between the external accumulation of OPEC and their capacity to import has been more of a blessing for the industrial countries than has been recognized.

Absorptive Capacity of Financial Institutions

A question often asked is whether the financial institutions of the world are capable of handling such large sums of OPEC surplus. The answer to this question requires an examination of the size of the financial markets of the industrial countries. Using the OECD data, it was estimated that the capital markets of the seven richest member countries absorbed around $450 billion in 1974. Thus the investable surplus of around $50 billion by OPEC amounts to 11 percent of the growth of flotation of new debt or stocks in these countries. Also, the total size of the financial markets in OECD countries was estimated to be close to $6,000 billion in 1974, which means that the surplus petrodollars amounted to less than 1 percent of the total. OPEC holdings are estimated to reach 2 or 3 percent of the total size of the capital markets by 1980.[13] So, there is no great danger of the failure of these markets to absorb such funds.

Two more comparisons may put the problem in a larger perspective. According to OECD, between 1975-85 the sum of $1,200-$1,600 billion is required for energy investment. OPEC surplus of between $150-$250 billion during this period will only amount to 10-15 percent of such investment. Also, a recent New York Stock Exchange report estimates that U.S. capital requirements in the same period will be $4,500 billion, the OPEC cumulative surplus will be less than 5 percent of the total.

The OPEC Balance Sheet

So far we have discussed the various myths surrounding petrodollar surplus and tried to bring out the realities of the situation. It was shown that although the problem could have been serious, a multitude of factors such as the rationality of OPEC spendings, the flexible exchange rates and the inherent efficiency and flexibility of the capital markets, prevented a financial crisis which many had expected. But where will the petrodollars go from here, how will

they be accumulated, and will the financial markets be able to resist
such strains in the future years? Fortunately most of the economists
from both the private industry and government have come to relatively similar conclusions. They all agree that all the previous forecasts were incorrect, that the surplus in trade balance will diminish
and the major portion of external accumulation of OPEC will gradually
disappear during the next decade. An interesting example of such
forecast is shown in Table 7.7. The following assumptions were
made for the forecast: (a) little change in demand for OPEC oil,
(b) an annual increase in OPEC per barrel "take" of 5 percent, (c)
an average increase in OPEC nonoil exports of 15 percent per annum,
(d) an average annual increase of 20 percent in the volume of OPEC
imports, (e) increases in the price of OPEC imports of 12 percent
in 1975 and 7 percent annually thereafter, and (f) financial return on
OPEC external investment of 8 percent per annum.

TABLE 7.7

The OPEC Balance Sheet, 1974-80
(billions of dollars)

	1974	1975	1976	1977	1978	1979	1980
Exports of goods and services	112	117	127	135	139	148	158
Oil revenues	105[a]	110	119	125	128	135	143
Nonoil exports	7	7	8	10	11	13	15
Imports of goods and services	50	65	83	108	138	177	227
Trade balance	62	52	44	27	1	-29	-69
Investment income	3	8	13	16	19	19	16
Current account	65	60	57	43	20	-10	-53
Grant aid	2	3	3	3	3	3	3
Surplus to be invested	63	57	54	40	17	-13	-56
External financial assets[b]	80	137	191	231	248	235	179

[a]Exceeded actual revenue receipts by $10-15 billion.
[b]Cumulative amount outstanding at year end.

Source: Morgan Guaranty Trust Company of New York, "World
Financial Markets," January 21, 1975.

It can be seen from Table 7.7 that the imports of OPEC countries will exceed their exports in 1979. However, the investment income of these countries will bolster the total accumulations to increase until 1979 when for the first time the cumulative surplus will decline. The 1980 cumulative surplus is estimated at $179 billion, compared to the import bill of $227 billion for OPEC in the same year. If nothing changes, the more developed OPEC countries will go into a deficit position and smaller states or the so-called "low absorbers" will maintain a small surplus throughout the 1980s. An interesting issue in the forecast is that of aid. It anticipates that aid will stabilize at $3 billion per annum during this period. In general, the forecast received powerful support from the deputy governor of the West German Central Bank Otmer Emminger and a U.S. Treasury official, Thomas Willet, who estimated the surplus at $200-$250 billion by 1980. Willet confirmed that his estimates represent the current appraisals of the situation within the Treasury and elsewhere in the U.S. government.[14]

Cooperation versus Confrontation

The increase in oil prices and the subsequent petrodollar surplus created a great deal of ill will in the West, with a huge press campaign and a great deal of political maneuvering. All that had gone wrong with the international economic system of the world was blamed on OPEC. The OPEC surplus of $50 billion or so was held to be the ruin of the industrial and developing countries of the world. But no one mentioned the fact that three countries (United States, Germany, and Japan) accumulated "nonoil" trade surplus of $58 billion in the same year.[15] Insofar as the poor and industrial deficit countries were concerned, such a surplus was exactly in the same ballpark as that of the producers, and just as damaging.

The United States, represented by Secretary of State Henry Kissinger and Secretary of Treasury William Simon, became the self-appointed leader of the world against OPEC. The U.S. motives for such leadership are unclear, since the United States was the least hurt by the increase in oil prices (over 60 percent of U.S. consumption of oil was domestically produced in 1974). Kissinger stated that he feared the economic recession in some European countries (notably Italy and France) might lead to a communist elector victory in these countries. Thus he sought to reduce oil prices by a variety of tactics (he assumed oil prices caused the recession). The outside observer cannot help but doubt his motives and reasoning. It may be noted that in many instances when the EEC countries and the Arab producers were nearing a dialogue on the petrodollar problem,

Kissinger dropped hints of military intervention or released films to the mass media of U.S. Marines doing desert exercises. Why did the United States adopt such an attitude? One can only suspect that those who saw the Arab influence and money as a direct threat to the state of Israel did their best to prevent Europe from getting too close to the Arabs--either on a global scale or through bilateral agreements.

The policy of confrontation was relatively unsuccessful, although it helped delay a mutual understanding. Many of the European countries made bilateral trade and cooperation agreements with the oil producers, while all the developing world stood solidly behind OPEC. The so-called military solution also lost its credibility. While civilians such as Robert Tucker and the State Department officials maintained that such an operation was feasible, the Pentagon dropped subtle hints that it viewed such an operation with skepticism because of its great operational difficulties and small chances of success. No help was expected from Europe.

The OPEC countries, particularly Iran and Saudi Arabia, kept their cool against militant demands by Algeria and Libya to raise the prices drastically or severely cut production to the levels sufficient for generating their revenue needs.

Insofar as the petrodollars were concerned, the industrial countries took an inherently inconsistent position. While they insisted that the surpluses be spent, they resented investments in their own countries with the cries of "the Arabs are coming!" Such an adoption of double standards particularly by the United States was treated with surprise and anger in OPEC. The United States has the largest single foreign investment in the world. The book value of her direct investments alone exceeded $100 billion in 1974.[16] After all, five of the seven major oil companies which discovered the OPEC oil are U.S.-owned. At the same time foreign investment in the United States is by no means meager. Some of the best known companies in the United States such as Shell, Lever Brothers, and the Nestle Company are foreign-owned. In 1973 alone, $13 billion of U.S. securities were bought by foreigners.[17] In 1974-75 several bills were sponsored by senators and supported by the State Department to limit foreign ownership of U.S. companies to 5-10 percent. None of these bills have yet become law, but they are proposed at a time when hundreds of U.S. companies hold equal or majority shares in joint ventures in the oil-producing countries. In the words of Assistant Secretary of Treasury Gerald Parsky,

> If the United States, with our historical support of
> free capital movements, were to adopt investment
> restrictions, other nations could be expected to take
> similar measures. At a time when the need for

worldwide cooperation is at a peak, the nations of
the world, led by the United States, would be re-
treating into isolated economic shells.[18]

Another major issue which is yet unresolved is the extent of
OPEC control over its recycled funds. Kissinger has consistently
opposed an international borrowing fund which obtains funds directly
from OPEC, since he believes that OPEC must be denied political
control over its money. A number of proposals have been put forward to ease the recycling process: IMF oil facility, the Kissinger
Plan, the Common Market plan, the Van Lennep Plan, and the Roosa
Plan.[19] But except in the case of IMF oil facility, OPEC was not
consulted in drawing up the plans.

To expect OPEC to give up political control of its funds is not
only unreasonable but also unworkable. OPEC does not wish to take
over the economies of the industrial nations, but rather act as an
institutional investor, with the ability to use this financial influence
to obtain cooperation in technical fields and preferential trade privileges (Iran, for example, is negotiating with EEC for preferential
access to the markets). Plans for OPEC to hand over its money to
mutual fund corporations to buy equities and government securities
around the world will simply not work. OPEC can easily hire its
own team of experts to place just as profitable investments with direct control over its funds.

Perhaps it is time for the United States to abandon its hostile
policy toward OPEC and reach a mutual agreement with it for future
recycling and investment policies--agreements that take account of
the interests of both sides and that may be relied upon as long-term
partnerships in the newly expanded club of the rich.

IRAN'S PETRODOLLARS: SURPLUS OR DEFICIT?

Iran as the second largest oil producer in OPEC was a major
beneficiary of the oil price hike. Iran's oil revenues rose 4.5 times
from about $4 billion in 1973 to $18 billion in 1974. As a result of
this change in the revenue the government substantially altered its
expenditure policy with a fully revised Fifth Plan, revised annual
budgets, larger defense expenditure, and aid and loans to both developing and industrial nations.

The Iranian forecast of oil revenues has determined the direction of Iran's investment policy at home and abroad. A large number
of commitments have been made for domestic investment, social
welfare, defense, and aid. The implementation of these projects require the realization of these revenues. In the following we will

consider the forecast of future revenues and expenditures in comparison with the likely events, and the alternative investment policies
which are open to Iran. Expenditures are considered in three broad
categories: internal expenditures, external expenditures, and defense. Internal expenditures include those specified in the Fifth Plan,
particularly those in oil, gas, and petrochemicals. External expenditures include investments in oil and nonoil activities as well as aid to
developed and developing countries.

Foreign Exchange Receipts

The balance of payments for the Fifth Plan (1972-73 to 1977-78)
forecasts total receipts of $114 billion (see Chapter 6). The oil sec-
tor is estimated to provide $102 billion or 90 percent of the total re-
ceipts. These figures are based on a production capacity and international demand for oil which are not likely to be realized. As a
result a considerable shortfall of revenues may occur which requires
a modification and reappraisal of the government objectives. Table
7.8 shows the details.
It can be seen that in both cases there will be a shortfall of 16
to 19 percent ($16.7-$19.5 billion) of the projected amount in the
Plan. Indeed the 1974-75 budget has been faced with a shortfall of
over $2 billion, with a possibly larger gap expected in 1975-76 budget.
But the last two years of the Plan, when the government had forecast
large increases in production, will create the biggest shortfall unless
oil prices are substantially increased. The cause of the shortfall is
twofold: a drop in the general level of demand for oil internationally,
and the inability of the Iranian oilfields to reach their forecast peak
capacity. At first sight one is tempted to conclude that there will be
no problems because Iran's balance of payments shows a surplus of
$17.5 billion (which was earmarked for aid), approximately equal to
the size of the deficit. The problem, of course, is not so simple.
Iran has made a great deal of aid and investment commitments which
cannot easily be discontinued. Furthermore, the gap will continue
to widen through the Sixth Plan (1977-78 to 1982-83) and adjustments
need to be made as soon as possible.
One important point deserves to be mentioned at this stage.
Iran's Fourth Development Plan envisaged expenditures which were
in excess of its projected revenues from oil; thus Iran's was an important force in raising the oil prices in 1971 to obtain the necessary
funds. A similar problem faces Iran in 1975, and it is quite likely
that the Shah will press for increases in oil prices of $1-$3 per barrel in the September 1975 meeting of OPEC to fill Iran's projected
foreign exchange shortage.

TABLE 7.8

Iran's Projected Receipts from the Petroleum Sector, 1973-80
(thousands of barrels per day and billions of dollars)

	1973	1974	1975	1976	1977	1978	1979	1980
Planned output[a]	5,314	5,676	6,322	6,958	7,596	7,603	7,606	7,610
Likely output[b]	5,860	6,030	5,500	6,000	6,200	6,500	6,500	6,500
Internal demand[c]	266	297	365	447	514	584	660	752
Stock change and loss	24	63	--	--	--	--	--	--
Exports	5,570	5,670	5,135	5,553	5,686	5,916	5,840	5,748
Revenue I[d]	3.8	18.0	18.9	20.5	21.0	21.8	21.5	21.2
Revenue II[e]	3.8	18.0	18.9	21.5	23.1	25.3	26.2	27.1

[a]Relates to Khuzistan production only. Other production was anticipated to be around 1 million b/d by the end of the decade.

[b]Based on technical and engineering capacities from 1976 onward.

[c]Based on 1974 projections by the Distribution Department of NIOC (unpublished).

[d]Based on a government take of $10.12 per barrel from 1975.

[e]Based on a 5 percent annual increase in government take from 1976.

Note: 1973 and 1974 data are actual.

Source: Compiled by the author.

The Fifth Plan envisages import of goods worth $79.1 billion.
Based on the 1972 import figure of $2.57 billion, this would imply a
growth rate in imports of just under 100 percent per annum. If defense expenditures of $29.1 billion (say with a foreign exchange com-
ponent of $20 billion) are included, then the anticipated growth rate
would be a little under 90 percent. If the present production level
and foreign exchange inflow is realized, then there would not be any
difficulty in the Fifth Plan; but even then, if the imports are allowed
to grow at these rates in the Sixth Plan, the most optimistic production forecasts cannot provide the necessary foreign exchange. If
Iran's imports are to rise to $40 billion in 1980 as predicted by Prime
Minister Hoveyda,[20] there would be no alternative but to borrow
from abroad to support this demand.

Let us first consider Iran's domestic expenditures. Details of
these expenditures in the Fifth Plan are discussed in more detail in
Chapter 6. In short, the government has undertaken to spend $69.6
billion in the five-year period. The public sector investment amounts
to $46.2 billion or 66 percent of the total, at an annual growth rate
of 38.1 percent compared to 14.1 percent in the Fourth Plan. These
include investments in capital goods, oil and gas, infrastructure and
social welfare, and the like. With the present level of government
commitment at home, it is highly unlikely that public spending be
cut down. It is thus reasonable to assume that domestic expenditures
would continue at their planned levels.

Domestic Hydrocarbon Investments

As a means of increasing its future export potential, the government has in the past few years concentrated its efforts on three
major hydrocarbon fields--domestic export refineries, petrochemicals, and gas.

Domestic Export Refineries

These types of refineries have for long been an attractive proposition to the producing countries, particularly Iran which has carefully been watching the export potentials of the Consortium-operated
Abadan export refinery. Export refineries provide two main advantages for the oil producers. It would provide them with the value
added involved in the refining process and a stronger political control on the supply. Because the refineries have since the late 1940s
been concentrated in the consuming areas, the consumer could switch
the source of crude in cases of difficulty (embargo or dearer prices),

but with the refineries in the producing areas the consumers will not
have such a flexibility.

Iran's drive to substitute products exports for crude started
with the takeover of the Abadan Refinery in 1973. Since then a number of preliminary agreements have been signed with Japanese, European, and American concerns. So far none of the agreements have
been finalized, because of lack of cooperation from consumers and
Iran's insistence, in some cases, on linking the refinery construction
to the construction of an adjoining petrochemical complex which can
use naphtha produced from the refinery as its feedstock. The Japanese group refused to submit to Iran's demand for an adjoining petrochemical plant (at an additional cost of around $1 billion), while the
German group's proposed 500,000 barrels per day refinery at
Bushehr is in danger of collapse because of the Common Market's
refusal to grant tariff privileges to oil products exports from Iran.
Without such privileges Iranian oil products will not be competitive
in Germany. Shell and a number of other companies have also reached
preliminary agreement on construction of an export refinery in Iran
and a joint venture in marketing in the U.S. east coast. But the agreement may not be finalized in its original form.

Although Iran has not succeeded in bringing refineries home so
far, the future trend is unmistakable. NIOC will seek partners which
have extensive marketing networks abroad and will provide incentives
such as discounts, lower fuel, costs (cheap natural gas), long-term
guarantees, and accessibility to capital. One can expect a general
movement in the direction of home refineries for Iran as well as for
other producers, but such a movement may be slower than has been
anticipated, particularly in view of the present excess capacity of refineries around the world. Iran's export of refined products is unlikely to exceed its crude exports given the lifespan of Iran's reserves
of oil and the state of refining operations around the world. Thus Iran
cannot look to export refineries as a major potential earner of foreign
exchange in the foreseeable future.

Table 7.9 shows the government's investment plans for oil. Investment in the oil industry in the Fifth Plan is projected at $9.3 bil-
lion--$5 billion from general revenue, $3 billion through NIOC, and
the remainder from the private sector. Refining absorbs the lion's
share of investment--44 percent of the total. Domestic refineries at
Isfahan, Tabriz (proposed), and Shiraz and Tehran II refineries constructed in the plan period,will not absorb more than half of the allocations. This shows the government's hopes for heavy investment in
port refineries which may well not be fully realized. Apart from
refineries, other investment commitments are expected to be realized
in full.

TABLE 7.9

Fixed Capital Formation for Oil During the Fifth Plan
(billions of rials)

Program	Public Sector	Private Sector	Total
Exploration and production	187.4	--	187.4
Refining	212.4	64.0	276.4
Transport and distribution	73.6	--	73.6
Nonbasic affairs	8.0	--	8.0
Operations outside Iran	30.7	--	30.7
Investment in affiliated companies	23.8	23.8	47.6
Total	535.9	87.8	623.7

Note: Investments in oil are not subject to credit limitations and may increase.

Source: Iran's Fifth Development Plan (revised), January 1975, p. 274.

Petrochemicals

Another important future source of income for Iran will be petrochemical exports. National Petrochemical Company (NPC), an NIOC subsidiary, was established in 1965, and has built up an industry with an investment value of $400 million comprising two wholly owned subsidiaries (Shahpour Chemical Company and Iran Fertilizer Company) and two joint ventures with U.S. companies. The joint ventures are Abadan Petrochemical Company (B. F. Goodrich 26 percent, NPC 74 percent) and Kharg Chemical Company (Indiana Standard 50 percent, NPC 50 percent). Table 7.10 shows the production of these ventures in the first quarter of 1974. Plans are under way to increase the capacity of these plants by 100 percent in three years. Three other petrochemical projects are now being implemented and a wide range of other projects are under discussion.[21]

The basic idea behind investment in petrochemicals is similar to that for export refineries--to capture the value added. But the difference is that the value added in petrochemicals is substantially above that of refining. At the same time export refineries are complementary to petrochemicals plants as the former's output is an important feedstock for the latter.

TABLE 7.10

Production of Iran's Petrochemicals
(first quarter of 1974)

Name of Company and Products	(Metric Tons)	
	Production	Sales
Shahpour Chemical Company		
Ammonia	39,950	25,664
Urea	29,728	28,161
Sulphur	69,639	44,381
Sulphuric acid	74,811	972
Phosphoric acid	23,532	--
Ammonium phosphate	53,496	55,319
Kharg Chemical Company		
Sulphur	68,845	80,900
Propane (barrels)	350,507	378,348
Butane (barrels)	208,620	210,123
Light naphtha (barrels)	208,191	207,259
Abadan Petrochemical Company		
Caustic soda	5,259	4,468
DDB	3,195	3,454
PVC	4,801	4,506
Polyca products	673	716
Iran Chemical Fertilizers Company		
Ammonium nitrate	8,243	8,258
Urea	13,558	11,794
Light sodium carbonate	1,173	160
Heavy sodium carbonate	6,691	2,263
Sodium bicarbonate	525	66

Source: Iran Oil Journal, August 1974.

Iran offers an attractive opportunity for petrochemical investments to foreign partners: cheap gas, security of feedstock, and a
possible geographical advantage due to its location between Europe
and Asia. Iran plans to spend $8 billion in a ten-year investment
program which may be the first step toward shifting the petrochemical
world's center of gravity to the Persian Gulf. By 1983 Iran aims to

be producing 5-10 percent of the world's basic petrochemical needs
and thereafter to supply 10 percent of the annual growth in the world
demand.[22]

Whether the investment figures quoted above can be realized or
not, it is clear that Iran is moving fast in the direction of petrochemical exports as the most economic way of utilizing oil. But to expect
petrochemicals to reach the above targets and provide a major source
of foreign exchange for Iran in the next decade or so is an optimistic
presumption.

Gas

At the time when Iran's oil production is leveling off, Iran's
gas industry is geared to assume a major role in the expansion of
the Iranian economy. Iran has ambitious plans to utilize its associated and nonassociated natural gas for domestic consumption, refinery and petrochemical feedstock, and gas reinjection to maintain
pressure at the oilfields and exports.

All of the gas produced in Iran is associated gas (produced with
oil and a free good in that sense), 58 percent of which was flared in
1973. A total of 4.8 billion cubic feet per day (cfd) were produced in
1973, of which 2 billion cfd were utilized. Gas consumption in the
same year for domestic, commercial, industrial, and petrochemical
uses amounted to 229 million cfd; oilfield consumption was 816 million cfd, and exports to the Soviet Union through IGAT (Iranian Gas
Trunkline) amounted to 840 million cfd. Exports to the Soviet Union
were expected to reach 1 billion cfd by the end of 1974 and then level
off (see Tables 7.11, 7.12, and 7.13).

The future export potential of gas and oil in Iran are dominated
by the key issue of reservoir reinjection for secondary recovery in
the Khuzistan oilfields (see Chapters 3 and 8). A look at the figures
will explain why. Estimates of the amount required for gas reinjection vary between 8 and 13 billion cfd, compared to the available
flared associated gas of 2 billion cfd by 1980. To fill this gap NIOC
is planning to use gas caps in suitable oil reservoirs, known gas
fields, and new gas discoveries which it is confident of making.[23]

Two important points have to be made at this point concerning
reservoir reinjection. First, reinjection is a costly venture involving collection of the gas, drilling reinjection wells, and reinjection.
Second, gas reinjection does not mean a loss of gas, but rather a delay in its utilization. One may expect around 85 percent of the rein-
jected gas to be recovered in 20 years or so, depending on the characteristics of the reservoirs.

TABLE 7.11

Associated Gas Production, 1973
(thousands of cubic feet)

Company	Gas Produced	Gas Utilized	Gas Flared	Percent Gas Utilized
NIOC/OSCO	1,615,683,300	724,292,941	891,390,359	44.8
IPAC	58,837,200	14,479,300	44,357,900	24.6
IMINOCO	27,057,000	1,000,000	26,057,000	3.7
LAPCO	26,692,774	1,457,500	25,235,274	5.5
SIRIP	10,991,900	1,157,500	9,834,400	10.5
NIOC (Naft-e-Shah)	4,316,900	463,100	3,853,800	10.7
Total	1,743,579,074	742,850,341	1,000,728,733	42.6

Note: This table overstates the amount of gas utilized, since some recipients (including IGAT and Kharg Chemical) flare a portion of the gas received.

Source: National Iranian Gas Company, in supplement to the Middle East Economic Survey, February 7, 1975.

TABLE 7.12

Associated Gas Production and Utilization, 1970-80
(millions of cubic feet daily)

	1970	1973	1975	1980
Production	2,772.0	4,784.0	4,949.0	5,793.0
Consumption				
Domestic, commercial				
Industrial and petrochemical	67.5	228.6	796.0	1,401.0
Oilfields consumption	924.0	816.4	1,164.0	1,367.0
Export	93.5	840.0	1,016.0	1,016.0
Total	1,085.0	1,885.0	2,976.0	3,784.0
Balance	1,687.0	2,899.0	1,973.0	2,009.0

Note: Export figures are for IGAT only. Balance indicated will be used for secondary recovery and other projects.

Source: National Iranian Gas Company, in supplement to the Middle East Economic Survey, February 7, 1975.

TABLE 7.13

Domestic Gas Consumption 1970-80 (Associated and Nonassociated)
(millions of cubic feet daily)

	1970	1973	1975	1980
Oilfield consumption	924.0	816.0	1,164.0	1,367.0
Domestic and consumption	0.3	1.2	4.0	42.8
Electricity generating	2.0	58.8	156.4	724.6
Industries	15.1	68.0	180.5	1,036.0
Total	941.4	944.0	1,504.9	3,170.4

Note: Does not include gas to be used for reinjection purposes. Industries heading includes direct reduction steel mill and petrochemicals.

Source: National Iranian Gas Company, in supplement to the Middle East Economic Survey, February 7, 1975.

Iran's gas exports are one of the government's hopes for the future. Although quite a few letters of intent have been signed, only two contracts were finalized. One is the exports to the Soviet Union which started a few years ago, and a major trilateral deal which was closed in April 1975. The trilateral deal involves Iran, the Soviet Union, and West Germany. The Soviet Union will receive 1.3 billion cfd of gas from Iran, and will deliver 1 billion annually to Germany over 25 years on account of Iran. The gas will be delivered in 1981 through a 900-mile pipeline which joins the southern gas fields to the Soviet border at Astara. At present prices this would mean an annual income of $250 million for Iran and $60 million for the Soviet Union. According to West German sources, parts of Iranian gas may be sold to Italy or Austria. [24]

Another serious contender for Iranian gas is El Paso Natural Gas of the United States and two Belgian companies, Sopex and Distrigaz, which have signed a letter of intent with the National Iranian Gas Company (NIGC), an NIOC subsidiary. The project involves LNG exports to Europe and the United States of 2 billion cfd initially, rising to 3 billion cfd from the early 1980s. [25] An earlier agreement with Transco of the United States for a $650 million project to produce natural gas liquids and methanol from Khuzistan was dropped by Iran in view of possible internal needs.

Iran's gas reserves are the largest in the Middle East and possibly the second largest in the world next to the Soviet Union. Earlier estimates put Iran's reserves at 270 trillion cubic feet or 65 percent of the Middle East reserves. [26] After the discovery of large gas deposits by Egoco of reportedly 175 trillion cubic feet, the estimates were revised upward. The Egoco discovery is said to be probably the largest gas field in the world. [27] Present estimates put Iran's reserves at 250-600 trillion cubic feet, half of which is located offshore. It is believed that the true magnitude is closer to the upper range. (Soviet gas reserves are estimated at 550 trillion cubic feet.)

As mentioned earlier, the key to Iran's gas export potential is the amount required for secondary recovery through reinjection. Domestic consumption, gas export projects, and gas required for reinjection would demand gas production of around 15-20 billion cfd by 1980, in comparison with the projected associated gas production of around 5 billion cfd in 1980. Thus the natural gas fields have to be tapped to supplement the associated gas produced as a by-product of oil. According to the Fifth Plan guidelines, $2.5 billion will be invested in the gas industry--$0.77 billion from government, $1 billion from NIGC, and the remainder from foreign investment. However, 57 percent of the total investment will be directed toward exports and 34 percent to refining and transport. No allocation is made for exploration, which indicates that sufficient resources are already known or that future exploration will be carried out by foreign companies. The total investment may well be insufficient to increase production to the desired levels. Thus, to ensure the availability of finance for gas, the Plan states that "the investment in the development of the gas industry will not be subject to financial limitations, and if necessary changes will be made in the projected credits."[28]

In an interesting paper Thomas Stauffer examines the potential value of gas in various uses such as gas reinjection, export refineries, urea, aluminum, iron, and LNG sales. His startling results have important policy implications for the Iranian gas industry.

TABLE 7.14

Value of Gas in Various Uses
(dollars per thousand cubic feet)

Gas reinjection	1.60
Refining	1.68
Urea	1.00-1.20
Aluminum	.94
Iron	.60
LNG	.89

Source: T. R. Stauffer, "Energy Intensive Industrialization in the Arabian/Persian Gulf," paper presented to Energy Seminar at Harvard University, April 1975 (unpublished).

Although the assumptions are oversimplified, the results are significant enough to warrant further investigations. The above results show that (a) energy-intensive industrialization may not be worthwhile for the Gulf countries, and (b) gas reinjection is economically sound and preferable to exports.

Export refineries, petrochemicals and gas offer attractive opportunities for Iran's future. The government is committed to large expenditures in these fields, and the likelihood of a sizable cut in these investments is remote. At the same time, these ventures have little export potential in the short term, though their long-term future will no doubt be bright.

EXTERNAL EXPENDITURES

In the following we shall consider Iran's foreign investments,
as well as aid and loan policy, and try to evaluate the importance
and the future trend of such policies.

Financial and Physical Assets

Unlike those of many other OPEC surplus countries, Iran's
investment in physical assets (such as real estate) is nonexistent.
There is also no indication of a movement of funds to buy government securities or unit trusts, or holding large sums of money in
bank deposits for investment purposes. Insofar as equities are concerned, there has been no rush for indiscriminate purchase of
equities, although Iran has bought shares in two companies, with a
third purchase in the pipeline. In July 1974 Iran bought 25.02 percent of the shares in the steelworks of Krupp Industries of Germany
for $100 million. In January 1975 Iran agreed to loan $245 million
to Pan-American Airways which provided Iran with an option to buy
13-15 percent of the stocks of the company (valued at $30 million at
current market prices). The Pan-Am deal also provides Iran with
a controlling interest in the airline's profitable Intercontinental
Hotels subsidiary for $55 million, bringing the total payments to $300
million. Another deal which was approved in April 1975 gives Iran
25.02 percent of shares in Deutsche Babcock & Wilcox AG, at a cost
of $75 million. The firm is a leading manufacturer of power gener-
ating equipment, produces nuclear power plants components, and is
active in a number of other industrial and engineering fields.
    Iran's purchase of equity shares abroad is governed by three
major criteria: effect on domestic economy (through technology
transfer and training), enhancing economic and political cooperation
between Iran and industrial countries, and the financial viability of
the project. The Krupp and Babcock purchases are directly related
to the future industrial needs of Iran in steel production and technology as well as nuclear power generation. The Pan-American
deal provides for training of staff and expansion of facilities of Iran
Air so that the airline may assume an international role. All the
deals have helped the economic and political cooperation between
Iran and host countries. Financial considerations are not always the
determining factor in such purchases. The Krupp and Babcock deals
were financially sound investments, although the Pan-Am purchase
was to some extent a rescue operation for an airline which had lost
money for six straight years and accumulated debts of over $335
million. Iran has made it clear that in no case would she wish to
control or influence the management.

In general such equity purchases have been relatively small in relation to Iran's petrodollar receipts. At the same time it is likely that such purchases in the future will be few, selective, and based on the above criteria.

HYDROCARBON INVESTMENT ABROAD

Iran's investment in oil and oil-related activities abroad have been relatively limited in the past. In the following we will consider exploration, transportation, refining, and marketing.

Exploration

NIOC went into partnership with British Petroleum a few years ago for exploration in the British sector of the North Sea, through its subsidiary Iranian Oil Company (U.K.). Very little information is available on the operation of this joint venture, though it is reported that no successful find has been made. In April 1975 NIOC won a concession to explore and exploit for oil in the continental shelf of Western Greenland. NIOC's fully owned subsidiary in Denmark has a 25 percent share in a consortium comprising British Petroleum, Standard Oil of California, and SAGA. The agreement covers an area of ten offshore blocks.[29] The soundness of investments in such "upstream operations" cannot be factually determined without more information, but for a national entity such as NIOC this may not be the most important factor. If NIOC wishes to gain experience through partnership with foreign partners, these joint ventures provide such opportunity.

Transportation

Iran's entry into international oil transportation started with the establishment of the National Iranian Tanker Company (NITC), an NIOC subsidiary, in 1955 (Iran has no pipeline ownership abroad). By 1974 the fleet included four oceangoing tankers, two with a capacity of 55,000 deadweight tons (dwt) each, and another two with a capacity of 35,000 dwt each. In 1973 two supertankers with a capacity of 230,000 dwt each were ordered from Japan, the first of which was delivered in April 1975. Iran has also agreed to buy a number of tankers from BP Tanker Company and pool them with the British Company under a unified (British) management late in 1975. Iran's ownership of tankers is very small in relation to Iran's export requirements and the bulk of Iranian crude will no doubt be exported

in the future through the traditional mediums. Furthermore, Iran
has no immediate use for the tankers bought recently. The Japanese-
made supertanker has been leased to a Japanese company, while the
BP Tanker Company will manage the Iranian tankers in the joint venture. In the future Iran may use its own tankers, particularly in view
of NIOC's increasing direct sales. In general, Iran's purchase of
tankers is a small limited operation and there is no indication of a
surge into the tanker business in the future.

Refining and Marketing

NIOC participates in two joint-venture refineries abroad, but
has no foreign distribution facilities. The two foreign refineries are
Madras refinery in India (operational in 1969), and Sassolburg refinery in South Africa (operational 1971). In both cases NIOC provided
expertise, part of the capital expenditure, and crude throughput on
a long-term basis (for details see Chapter 8). In July 1971 NIOC
acquired 24.5 percent of the shares of the Madras Fertilizer Plant,
the largest and most modern fertilizer complex in India.

After the February 1971 Tehran Agreement, when oil prices
started to rise for the first time after 11 years, NIOC undertook a
series of studies for expanding its refining and marketing operations
in Europe and the United States. A number of letters of intent were
signed with Greek, German, U.S., and Belgian concerns. The
Greek deal involved participation in ownership of Asperoprygos refinery and marketing of its products. The German agreement entailed participation of NIOC in a number of refineries in Germany as
well as equity participation in 1,300 service stations.[30] The U.S.
deal provided for participation of NIOC in 50 percent equity ownership of Ashland Oil Company. Ashland has a 60,000 b/d refinery
and a petrochemical complex in Buffalo as well as 180 service stations in New York State.[31] Another U.S. deal between Shell and
other U.S. companies for construction of a 500,000 b/d refinery in
Iran and a joint marketing venture in service stations in the U.S.
east coast was signed in 1974, but was postponed indefinitely in
1975.[32] The most dramatic event, however, was the failure of the
proposed $200 million Irano-Belgian refinery. Although the Belgians
dragged their feet in giving the final approval, NIOC's telegram of
cancellation on the ground that "the conditions for the deal are no
longer valid" came as a great surprise to many experts.[33] Despite
the readiness of the Belgian Prime Minister and six other cabinet
ministers to fly to Tehran immediately to salvage the deal, the
Iranians washed their hands of it. As a result, the Belgian Government fell.

All the post-1971 agreements had two things in common. First, NIOC was required to provide crude on a long-term basis and at preferential prices. Second, all the deals were unsuccessful. Although the exact details of the deals are not available, there are indications to the effect that in most cases NIOC did not find the project attractive enough to commit itself to long-term supplies and price guarantees. But what makes such investments attractive, and has NIOC adopted a rational policy toward foreign projects? These questions will be discussed below.

THE CASE FOR AND AGAINST
DOWNSTREAM OPERATIONS

The term <u>downstream operations</u> signifies the process through which exploited oil (upstream operations) is marketed. The process involves transport, refining, and distribution. The economic rationale for downstream operations in the consumer countries has been the subject of great debate among the oil producers in the past decade or so. A number of emotional and mistaken conclusions were arrived at, mostly on the false premise that all the profits in the oil business are at the downstream level. This in fact is not true. The misunderstanding arises from the notion that the oil companies make excessive profits, which can be reaped by the oil producers in downstream ventures.

Before we go any further, let us determine how much money has in fact been made by the oil companies. It is true that the internal accounting practices enable the integrated oil companies to assign fictitious transfer prices to various stages of their operation in order to minimize their tax liability to various governments (except for oil producers, since taxes are not based on profits in OPEC). But what these companies cannot do is to manipulate the end product or the final net income, which is published every year in their annual reports for the benefit of their shareholders. Two other points have to be emphasized here: first, it is possible that to maintain market share in the long run they accept lower profits in the short run as a matter of corporate strategy; second, they have diverse interests in other nonoil energy materials such as coal, shale oil, uranium, and power generation, which reduces their profit level in the short term. However, if we consider the rate of return of these companies over a long period of time, none of the above two factors become relevant, since they can be interpreted as short-term sacrifices for long-term gains. Since a slice of the cake cannot be larger than the cake itself, it is reasonable to assume that downstream profits are smaller or at most equal to the net income on total operations (including crude

production). Thus we can in fact take the final net income figures
from the annual reports as representative of downstream profits.

Profit figures are on their own meaningless. We must know
the investment upon which the profit was generated. Thus the rate
of return on investment (ROI) which is the ratio of net income to the
stockholders' equity is a much more meaningful concept. To be
sure, there are enough profits in the downstream operations to make
them financially attractive for the oil companies, but the rate of return of the oil companies was a normal rate of return of 8-14 percent
per annum in the 1960s and early 1970s. This normal rate of return
is in line with the return from other manufacturing industries. The
year 1972 was a particularly difficult one for the oil companies, because of the generally depressed condition of the oil industry. British Petroleum's net income fell by 59 percent, Gulf by 65 percent,
and Shell by 22 percent; the ROI (return on investment) was between
10-12 percent for most of the companies. The year 1973 was an ex-
ceptionally good year for the oil companies, because of the increase
in demand, increase in prices, and inventory profits. Their earnings were also boosted by currency devaluations. Exxon's net income rose by 59 percent, BP's by 363 percent, Shell by 159 percent,
Gulf by 306 percent, and other majors between 45-90 percent. Nevertheless, despite these increases in income, their rate of return
did not rise as dramatically. For instance, Exxon's ROI was 19 percent, BP's 16 percent, and Gulf's 14 percent. These returns were
indeed above the normal ROI, but not by the magnitudes expressed
in terms of profits. The year 1974 again was a profitable one for
the oil companies, though first quarter figures for 1975 suggest that
four of the five major U.S. companies suffered a loss in net income.[34]
Thus, profitability and ROI must be considered on a long-term basis
rather than short-term spurts, and on that basis the return has not
been more than normal in the past decade.

Another way of expressing the same problem is to consider
the per barrel profit of the oil companies. The profit per barrel
for the companies has oscillated between 30-80 cents in the 1960s
and early 1970s to around 80-120 cents in 1973 and 1974. This compares to the Middle East Government's take of 80 cents in 1970 and
$10 in 1975. This means that before the crude prices began to in-
crease, what could have been earned at the downstream level was
substantial compared to what the governments were earning. With
the present price levels this prospect is no longer attractive. In
other words, at present crude sales is a much more profitable business than marketing the products.

There is no gold at the end of the rainbow, and indeed, the
author remains unconvinced of the financial attraction of such projects.
However, these conclusions are based on the assumption that OPEC

will hold together and will keep crude sales the most profitable part
of the operation. If the oil producers are not unsure of such solidarity, then some downstream activity may become necessary. Furthermore, if the foreign operations are offered as part of a package deal
which involves domestic export refineries and petrochemical plants,
such deals may be worthwhile. Also, very limited operations abroad
may be justifiable for oil producers as a matter of strategy--to learn
about the oil industry in consumer countries.

Insofar as NIOC is concerned, such marketing operations are
even less attractive: Large investments are required for such operations, NIOC will have to streamline its managerial talent for running such ventures and,more important, Iran's proven reserves are
not large enough to warrant such long-term investment policy. All
such joint ventures are dependent on long-term supply contracts and
guaranteed prices--commitments NIOC would be wise not to enter
into except for special circumstances referred to above.

AID TO DEVELOPING COUNTRIES

Iran's aid to developing countries prior to 1974 was limited to
a $14 million loan for an irrigation project to Morocco in 1967 and
$5 million to Pakistan, Senegal, Jordan, and Tunisia in 1973. As of
1974 Iran entered the big aid league with a multibillion dollar aid program for the developing nations. Detailed information on such aid is
not available, but recent reports indicate that provisional aid figures
shown in Table 7.4 may be an underestimate. According to Mehdi
Ghaffar-Zadeh, Iran's Vice Minister for Industry, Iran's total commitments in the first 11 months of 1974 amounted to $7.7 billion,
60 percent of which went to developing countries.[35]

Iran's commitments to multilateral organizations include:
$2.5 million to UN agencies, $20 million to UN Emergency Fund,
$150 million to OPEC Fund, $200 million in World Bank bonds, and
$700 million to IMF oil facility. Although no details on the actual
disbursements are available, it is understood that except for the contribution to OPEC Fund, the remainder are fully utilized.

Iran's bilateral aid is shown below to indicate the characteristics of such payments.

Egypt: Iran's largest aid agreement involves $1 billion of as-
sistance to Egypt. It involves $400 million aid contribution to re-
vitalize the Egyptian industry; $250 million for the reconstruction of
Port Said and the Suez Canal zone. This loan carries a low interest
rate and involves the assistance of Iranian technicians over a long
period of time. $100 million to finance Egypt's purchase of Iranian
made buses, road-building equipment, and machine tools; $100 million

for imports; $120 million for financing another oil pipeline along the Suez Canal and establishment of petrochemical industries along the pipeline as well as widening the canal itself. Also, a joint investment bank in Cairo with a capital of $20 million will be set up.

India: The second largest recipient of Iran's aid is India, with a commitment of $900 million. The aid provides for shipment of $500 million of crude oil to India over five years at 2.5 percent interest with a grant element of 40 percent. Iran is also believed to have sold India oil at concessional rate under an agreement in February 1974. In August 1974 Iran agreed to provide $400 million as an untied credit to India. Other projects include iron ore, refinery expansion, joint shipping line, and petrochemicals.

Pakistan: Iran's third largest aid commitments are to Pakistan for $643 million. It includes a loan of $580 million committed in July 1974 to help meet Pakistan's balance of payment and development needs. The loan will carry a 2.5 percent interest and a grant element of 30 percent. In August 1974 Iran offered Pakistan a $63 million loan for regional development. Concessional sale of oil to Pakistan is also reported.

Syria: $150 million, low interest rates to finance purchase of industrial goods.

Sri Lanka: $67 million, advanced payments for Sri Lankan imports and credits for a proposed $150 million fertilizer plant.

Bangladesh: $15 million, loans repayable on easy terms over a long period for establishment of a fertilizer plant. In January 1975 Iran agreed to sell Bangladesh 300,000 tons of crude at "very favorable terms."

Jordan: $13 million, interest-free, to finance housing project for Jordanian officers.

Afghanistan: $10 million, easy terms over long period for agriculture and industrial ventures.

Senegal: $16 million, repayable at 2.5 percent interest over 20 years for irrigation and industrial projects. The credit also involves a joint refining project.

Tunisia: $8.5 million, similar to Senegal's terms but no specific projects.

Other aid: Morocco, Sudan, Lesotho, and Zaire (unspecified). Also $1 million to Sahel for victims of the drought and $1 million to Pakistan for earthquake victims.[36]

LOANS AND TRADE WITH INDUSTRIAL COUNTRIES

In a bid to obtain technological cooperation and show her goodwill to industrial nations, Iran has engaged in a number of loan and trade deals with these countries. The major deals involve the United Kingdom, France, Italy, and the United States.

United Kingdom: A loan of $1.2 billion carrying commercial interest rates and guaranteed by the British government. The loan will open lines of credit to public sector industries, will be given over three years, and is repayable after five years from the entry into effect of each part of the loan. The sum of $400 million has already been disbursed.

Italy: A series of preliminary agreements have been signed with Italy aiming at a $3 billion trade and cooperation between the two countries. They include: a steel mill based on direct reduction method with a capacity of 3 million tons a year at Bandar Abbas, a satellite town with 80,000 inhabitants at Bandar Abbas, shipbuilding facilities with a capacity of 750,000 tons per annum, another tire factory by Pirelli with a capacity of 400,000 tons a year, NIOC-ENI joint venture for refining and distribution of oil products in Europe and Africa. Other agreements include: petrochemicals, aluminum, textiles, construction, agriculture, and capital goods manufacture. No actual disbursements have taken place.

France: A series of preliminary and final agreements worth $7.8 billion, including $1.8 billion for the two French atomic power plants. Other projects are: 41-mile underground subway for Tehran, production of 100,000 Renault cars in Tran as a joint venture, construction of 200,000 housing units, color TV system for Iran based on the French "Secam" system, 26 turbo trains as well as electrification and modernization of the Iranian rail network, a fleet of methane tankers, a $270 million steel plant. Other projects include cooperation in agriculture, petrochemicals, and energy. The sum of $1 billion has actually been disbursed.

United States: The largest ever single agreement between two countries was signed by Iran and the United States in March 1975 involving $15 billion over five years. Five billion dollars will be in normal trade items representing an increase of some 20 percent per year in nonoil trade; another $5 billion is for the sale of U.S. military equipment to Iran and another $5 billion in actual U.S. involvement in the Iranian economy. They include 8 nuclear plants of 1,000 megawatts each, associated desalination plants, 20 prefabricated factories, 100,000 apartments, 5 hospitals with a total of 3,000 beds, the establishment of an integrated electronics industry, the building of a major port, joint ventures to produce fertilizers, pesticides, farm machinery, and processed foods, superhighways, and vocational centers. No actual disbursements.

It is important to note that based on the information available, these agreements have not included a direct oil for technology exchange, though the oil component may indeed be hidden among various clauses. Except for the U.K. and U.S. deals, no time limits have been set, and indeed so far only $1.4 billion of the total

committed $27 billion has been disbursed. Disbursement of the re-
mainder will, no doubt, depend on Iran's foreign exchange position
in the next few years.[37]

DEFENSE EXPENDITURES

Iran's defense expenditures have been the largest single foreign exchange receiver. Traditionally, defense expenditures have
accounted for between 25 to 30 percent of the total government disbursements. In 1962 defense and security expenditures amounted to
$210 million, in 1966 $360 million, in 1970 $820 million, and in 1972
$1.4 billion. After the increase in oil prices in 1973, the govern-
ment general budget increased from $4.8 billion in 1972 to $23.7
billion in 1974-75 and to $26.5 billion in the proposed 1975-76 budget.
However, the share of defense and security expenditures remained
surprisingly stable at $5.3 billion and $7.8 billion for the last two
budgets.[38] The Fifth Plan anticipates total government receipts of
$122.8 billion, of which $29.1 billion or 24 percent will be spent on
defense.[39] Iranian data do not show the foreign exchange component
of such expenditures (imports) but it is reported that between 50-80
percent of such expenditures are imports. Table 7.15 shows the expenditures in the Middle East on defense.

Iran's defense expenditures have placed a heavy burden on
Iran's resources. The increase in oil revenues and prices has made
the Persian Gulf region an increasingly important lifeline for the
West. Iran intends to ensure the safety of this lifeline and to main-
tain an effective sphere of influence in the Persian Gulf and Indian
Ocean, with the cooperation of the neighboring countries.

Iran's arms purchases have no doubt triggered similar purchases in the neighboring countries. Saudi Arabia and the United
Arab Emirates, for instance, have the highest per capita arms purchase. However, insofar as military power is concerned absolute
magnitudes of arms are a more meaningful proxy for military
strength rather than percentage of GNP or per capita, and Iran is
easily the largest consumer of arms in the Middle East.

SUMMARY AND CONCLUSION

In the course of this chapter two specific issues were discussed: the petrodollar problem on a global scale and the particular case of Iran with alternative policies for utilization of the newly
found oil wealth.

TABLE 7.15

Defense Expenditures in Middle East, 1974

Country	Defense Expenditure (billions of dollars)	Per Capita (dollars)	Percent of GNP
Algeria	0.404	25	3.3
Bahrain	0.008	35	2.2
Egypt	3.117	85	35.8
Iraq	0.803	76	7.0
Jordan	0.142	54	10.9
Kuwait	0.162	154	1.5
Lebanon	0.133	42	4.3
Libya	0.402	178	3.5
Morocco	0.190	11	3.0
Oman	0.169	228	15.4
Qatar	0.023	130	1.1
Saudi Arabia	1.808	228	5.7
Sudan	0.118	7	5.9
Syria	0.460	65	16.4
Tunisia	0.043	8	1.5
UAE	0.140	821	2.8
North Yemen	0.058	9	3.6
South Yemen	0.029	18	4.8
Total 18 Arab States	8.209	60	7.1
Israel	3.688	1,131	42.4
Iran	5.328	165	13.3
United States	85.500	400	6.2

Source: Middle East Economic Survey, February 21, 1975.

The petrodollar problem has been exaggerated out of proportion by hasty doomsday predictions as well as political interest groups
in the West. The anticipated surpluses and the financial collapse did
not come true, the money did not unproportionally flow to specific
countries, and the capital markets managed to deal efficiently with
the funds. Indeed, the most recent forecasts point to a deficit in the
trade balance of these countries by the end of the decade and a total
accumulated surplus of $179 billion by 1980. Although the developing
countries suffered from the increase in oil prices, they found the
OPEC countries generous aid donors in comparison to the industrial
countries of the West.

Insofar as Iran is concerned, it was shown that the country's future foreign exchange position may well be a cause for concern. Although the present foreign exchange reserves of around $8 billion will ensure that Iran is in no immediate danger, adjustments will have to be made soon if Iran is to avoid becoming a deficit country in a few years' time.

Future export earning hopes on gas, petrochemical, and export refineries are misplaced. To be sure, these industries may have a bright long-term future, but in the short term Iran cannot expect substantial earnings from these sources.

Iran has correctly curtailed its downstream activities abroad. There is no great financial advantage to be gained from these ventures, except possibly as a matter of corporate strategy on a very small scale. Iran's proposed investment in foreign oil ventures in the Fifth Plan amounts to $460 million, indicating 5 percent of the total investments in the five-year period.

Iran's aid to developing countries has been generous, though the direction of aid has been clearly linked to the future Iranian political influence in the region. A great deal of humanitarian aid has also been granted to the African nations. Egypt as the leader of the Arab world is the largest recipient of aid from Iran, India the giant nation with nuclear capabilities is in the second place, and Pakistan-- Iran's CENTO ally--in the third place. Committed aid to these three countries accounts for 85 percent of the promised aid by Iran in 1974. Despite the tensions between India, Pakistan, and Afghanistan, Iran has managed to hold a delicate balance in between. Trade and loans to the industrial countries have also been on a substantial scale with the particularly worst hit countries such as Italy and the United Kingdom as well as the United States and France. The total proposed deals with the industrial countries amount to $27 billion, 5 percent of which has already been disbursed.

Defense expenditures by Iran have increased at a surprisingly high rate from $1.4 billion in 1972 to $7.8 billion in 1975-76. Iran was the largest purchaser of arms in the Middle East in 1974, but from such a position of strength, Iran managed to mend her relations with her long-time foe Iraq. The Shah has made it clear that he has no intention of occupying territories in the Gulf; rather he would wish the local governments in the region to replace the U.S. and Soviet presence in the Persian Gulf and the Indian Ocean.

One thing is certain, Iran will not be able to keep up expenditure at home, aid and loan, and defense at the present levels of 1974 and 1975. Iran expects a $2 billion return on its proposed $11 billion investment abroad during the Fifth Plan, but this will not be sufficient to maintain the present level of expenditures. From the three major components referred to above, aid and foreign investment are the

first to go. Indeed, this has been made an official government policy.
Defense may also be cut down before any large-scale reductions in
domestic expenditures are undertaken. Indeed, the recent $5 billion
arms deal with the United States for a period of five years may well
indicate that defense expenditures will slowly fall in the future.

Whether the domestic expenditure should be reduced or the
government should go back to the capital markets to borrow funds is
a matter beyond the scope of this chapter. However, a few general
observations may be made: the inflow of imports in 1974 was about
$10 billion compared to $3.6 billion in 1973. The imports included
a large component of consumer goods and food items which can be
reduced to ease the foreign exchange shortfall. This can be done by
raising the tariff barriers back to the previous levels in 1974 (in 1975
tariffs were substantially reduced). Only food import subsidies in
1974 amounted to over $1 billion. The infrastructure is already show-
ing signs of strain. The Fifth Plan forecasts a shortage of over
720,000 workers; only 10,000 of the shortage is in the unskilled
category, the bulk of the shortage is highly skilled and semiskilled
labor which Iran lacks. In 1974 alone Iran had to pay $100 million
in fines to the shipowners for delays in unloading of the cargo in the
Iranian ports. Custom posts are jammed, and the government does
not have enough warehouses to store the imported wheat.

The government of Iran is clearly interested in distributing the
newly found affluence to the bulk of the populace. It hopes to maintain this affluence through a dynamic free-enterprise economy, with
the help of foreign technology and labor. But maybe the injection of
capital and imported goods into the economy is too fast, for the infrastructure and the economy as a whole. If Iran is to experience a
shortfall in foreign exchange requirements with a direct impact on
the standard of living of the people (such as cutting food import subsidy), the nation will find it more difficult to adjust itself to a lower
standard of living after a few years of affluence. Perhaps the government should consider maintaining its development goals, but implementing them at a slower pace--keeping in mind the future position of Iran's foreign exchange.

NOTES

1. Morgan Guaranty Trust Company of New York, "World
Financial Markets: The Build-Up of OPEC Funds," September 23,
1974.
2. Ibid.
3. Walter J. Levy, "World Cooperation or International
Chaos," Foreign Affairs 52, no. 4 (July 1974); Gerald R. Pollack,

"The Economic Consequences of the Energy Crises," Foreign Affairs 52, no. 3 (April 1974); Thomas R. Stauffer, "Oil Money and World Money: Conflict or Confluence?" Science 184, no. 19 (April 1974).

4. For a refutation of these myths see F. Fesharaki, "The Petrodollar Myth," Kayhan International, January 12 and January 18, 1975; J. Amuzegar, "Reassessing the Effects of the Oil Price Hike," Kayhan International, February 22, 1975.

5. William Simon, testimony before the Subcommittee on Financial Markets of the Senate Finance Committee on January 30, 1975.

6. Ibid.

7. Ibid.

8. For an interesting survey see H. G. Johnson, Economic Policies Towards Less Developed Countries (London: Allen & Unwin, 1967).

9. See OECD, Flow of Resources to Developing Countries (Paris, 1973).

10. J. Hewson and E. Sakakibara, The Eurocurrency Markets and Their Implications (Boston: Lexington Books, 1975), chap. 7.

11. For a good series of analyses see J. N. Bhagwati and S. Eckaus, eds., Foreign Aid (London: Penguin Modern Economics, 1970).

12. IMF Survey, November 18, 1974.

13. Stauffer, op. cit.

14. Middle East Economic Survey, January 31, 1975.

15. The Economist, March 22, 1975, p. 87.

16. Gerald Parsky, testimony before the Subcommittee on Multinational Corporations of the Senate Committee on Foreign Relations, March 18, 1975.

17. Ibid.

18. Ibid.

19. For a brief summary see Newsweek, January 16, 1975, pp. 26-27. The Roosa Plan appeared in Foreign Affairs 53, no. 2 (January 1975).

20. Reported in Kayhan International, April 12, 1975.

21. Middle East Economic Survey (Supplement), "Iran: Gas on the Move," February 7, 1975.

22. European Chemical News, June 7, 1975.

23. Middle East Economic Survey, op. cit.

24. Kayhan International, April 19, 1975.

25. Middle East Economic Survey, January 3, 1975.

26. Economist Intelligence Unit, "Oil Production, Revenues and Economic Development," QER Special No. 13, 1974, p. 9.

27. Oil and Gas Journal, July 15, 1974, p. 29.

28. Iran's Fifth Development Plan, Plan and Budget Organization, January 1975, p. 279.

29. Kayhan International, April 26, 1975.

30. Kayhan International, December 8, 1973 (Persian language edition).

31. Petroleum Press Service, September 1973.

32. Petroleum Intelligence Weekly, April 28, 1975.

33. Financial Times, January 17, 1974, and Sunday Times, January 20, 1975.

34. For further details see the annual report of these companies and Petroleum Press Service, May 1973 and May 1974. Also, see reports by Chase Manhattan Bank and First National Bank on profitability of the oil companies. Note: the latter two sources must be read with caution in view of their connections with the major oil companies.

35. Kayhan International, December 7, 1974.

36. The information was collected from the World Bank, IMF Surveys, and Kayhan International.

37. Based on reports in Kayhan International.

38. The Iranian annual budgets, 1962-75.

39. The Fifth Development Plan of Iran, op. cit., p. 41.

Part III is basically concerned with the activities of the National
Iranian Oil Company (NIOC) in the domestic oil market. It discusses
the production, distribution, sales, and pricing of oil products throughout the country, and thus the Constitution and structure of NIOC are
clearly relevant in this connection. Although the book is concerned
with economic rather than legal and organizational aspects of NIOC,
the latter are of significance and form the background to the investigations in the following chapters.

After the Iranian nationalization, NIOC was created to run the
domestic oil industry. The first NIOC Constitution was approved by
Parliament in 1955, the second in 1968, and the third in 1974. The
1974 Constitution was intended to supplement rather than replace the
previous versions; thus reference will be made to both 1968 and 1974
Constitutions in the following text.

THE NATURE OF THE COMPANY

Contrary to what might be implied by the name of the Company,
NIOC is not a public corporation. It is, according to its Constitution, a commercial company, paying tax on its net profit at the rate
of 50 percent. The Company has 10,000 shares of 1 million rials
each, representing a share capital of about $150 million. The paid-
up capital constitutes 50 percent of the nominal share price, and the
unpaid capital could be called upon at any time, subject to the approval of the Cabinet.[1] All the shares are held by the Iranian government and are not transferable. An increase in the issue of
shares would have to be requested by the Board of Directors and
approved by the Cabinet.

The assets of the Company include the ownership of all petroleum and natural gas resources of the country, as well as pipelines,
bulk depots, road tankers, and filling stations necessary for the distribution of petroleum products. The assets also comprise all the
buildings, installations, and fixed equipment operated by the Consortium in the Agreement Area.[2]

DUTIES AND LEGAL POWERS OF NIOC

The 1955 Constitution vested the ownership of all equipment
classified as fixed assets and operated by the Consortium, in NIOC.

It granted the Company general administration of all oil and natural gas resources and the unrestricted right to search for oil, produce, refine, and distribute petroleum products throughout the country. The Company was given a statutory monopoly of domestic sales. Moreover, the Company was charged with the responsibility of operating the nonbasic activities in the Consortium Agreement Area.[3]

The Constitution does not limit NIOC's activities exclusively to oil operations. Thus, for example, the Company may build pipelines, own tankers, operate a distribution network both at home and abroad.[4]

The 1957 Petroleum Act provided NIOC with legal powers to grant permission to any Iranian and/or foreign company to take part in the exploration and production of crude oil, either in partnership with NIOC or on their own.[5] It also empowered the Company to form subsidiaries in Iran and abroad, but all the shares of the Company are to be held by NIOC.[6] Moreover, NIOC is exempted from import licensing for all its foreign equipment and pays no duty on such imports (Article 8 - 1968).

THE BOARD OF DIRECTORS

NIOC's Board of Directors in effect runs the Iranian oil industry, and its Chairman, who is also the Managing Director of the Company, unofficially holds the status of a superminister. Until 1962, Mr. A. Entezan was the Chairman and Managing Director of the Company, but since 1962 Dr. Eqbal, a former Prime Minister, has been occupying this position. Dr. Eqbal was appointed by the Shah and reports directly to him, which indicates the importance attached to the leadership of NIOC by the government.

The 1974 Constitution raised the number of board members to ten, including the Managing Director. There are also five alternate directors selected by the board. All the board members must have at least ten years of experience in the oil industry and must have occupied important positions for five years. In the new board the heads of two main NIOC subsidiaries, the National Iranian Petrochemical Company and the National Iranian Gas Company, are included. This reflects the newly found importance of natural gas and petrochemicals in the overall export policy of the company, and the need to involve the heads of these companies in the overall policy-making machinery of NIOC.

Almost all matters regarding the administration and policy of the Company are decided by the Board of Directors. The only restriction on NIOC decision making is with regard to the pricing of the four main products: gasoline, kerosene, gas oil, and fuel oil.

Any such increase in prices must be recommended by the Board of
Directors to the Cabinet for approval. Only after such approval can
prices be altered (Article 34b - 1974).

NIOC is, contrary to its Constitution, in all respects a public
corporation. It has no profit maximization motive and its operations
comply with the overall government policy rather than the making of
profits. The organizational structure of the Company has undergone
some change, if only in terms of increasing the complexity of the
Company. There has certainly been a tendency to decentralize the
decision making of the Company, but this decentralization may not
have gone far enough. The Company is under close scrutiny by the
government and the presence of cabinet ministers ensures that its
wishes are carried out. Indeed, the control of prices of the four
main products by the government has effectively placed the control
of the most important source of NIOC revenue in the hands of the
government. In other fields, such as production and refining, NIOC
has been obliged to ask for the government's approval before it could
put its projects into operation.

NOTES

1. The representatives of shareholders at the General Meet-
ing of the Company are the Prime Minister, Minister of Finance and
Economics, Minister of Energy, Minister of Industry and Mines,
Minister of Labor and Social Welfare, Head of the Plan and Budget
Organization, and another minister to be decided by the premier.

2. For details see NIOC Constitution, Articles 1, 2, 3 - 1968
(in Persian).

3. Nonbasic operations include health, education, recreation,
housing, and general welfare of the employees.

4. NIOC Constitution, Article 4 - 1968.

5. The 1974 Constitution restricts ownership of NIOC to at
least 50 percent of the joint venture.

6. NIOC has three subsidiaries at present--the National
Iranian Gas Company, the National Iranian Petrochemical Company,
and the National Iranian Tanker Company.

CHAPTER

8

SUPPLY I—PRODUCTION, REFINING, AND EXPORTS

Iran has for long been a leading oil producer in the world. Until 1951 she was the largest oil producer outside the Soviet bloc and the United States. In 1951, Iran supplied 2.9 percent of the world production and 18.9 percent of the Middle East oil (see Table 8.1). After the Iranian nationalization the major oil companies expanded their activities in the other producing countries and thus the Iranian lead was lost. In 1960 Iran was the fourth largest producer out of the five OPEC members, but since 1971 she has been the largest producer (and exporter) after Saudi Arabia. In 1973 Iran supplied 28 percent of the Middle East production, 10.5 percent of the world production, and 21 percent of the world exports.[1]

THE IRANIAN CRUDE PRODUCTION

The search for oil in Iran is undertaken by various operators: (a) the National Iranian Oil Company, (b) the Iranian Oil Operating Companies (the Consortium),[2] (c) 11 groups of companies involved in the joint venture agreements: SIRIP, IPAC, IROPCO, IMINCO, LAPCO, FPC, PEUPCO, DOPCO, INEPCO, BUSHCO, and HOPECO. There are also foreign companies drilling for oil under service contracts: SOFIRAN, PHILIRAN, EGOCO, DEMINEX, TOTAL, ULTRAMAR, AGIP, and Ashland.[3]

In terms of production, the Iranian scene is dominated by the Consortium. The National Iranian Oil Company produces oil from one oilfield in Nafte-Shah and the size of its production is small. The service contracts have not yet produced any oil in commercial quantities, but four of the joint ventures are producing crude. In all years the Consortium production has accounted for over 90 percent of the Iranian production. NIOC production has increased 2.7 times

202

during this period, but in 1974 it accounted for only 0.3 percent of the total. SIRIP, the first Iranian joint venture, has been relatively unsuccessful. It took four years to start production and it produces the smallest quantity of oil compared to the other partners. LAPCO has been the most successful nonconcessionary venture; within the span of four years it became the largest producer after the Consortium.

TABLE 8.1

Iranian Crude Production in Comparison with OPEC,
Middle East, and the World in Selected Years

Year	Thousands of Barrels per Day				Relative Importance of Iran, Percent		
	World	Middle East	OPEC	Iran	World	Middle East	OPEC
1951	12,290	1,920	--	355	2.9	18.5	--
1955	16,190	3,220	--	329	2.0	10.2	--
1960	21,046	5,269	7,876	1,052	5.0	20.0	13.3
1964	28,360	7,427	12,070	1,690	5.9	22.8	14.0
1968	38,835	11,170	17,652	2,848	7.3	25.5	16.1
1971	48,675	16,065	25,083	4,540	9.3	28.3	18.1
1973	55,970	20,895	30,992	5,861	10.5	28.0	18.9

Note: Middle East production excludes production in Turkey, Syria, and Israel.

Source: Compiled from NIOC publications, BP Statistical Review of the World Oil Industry, and OPEC Annual Statistical Bulletin.

IRAN'S OIL EXPORTS

Iran's oil exports are dominated by the Consortium. The percentage share of the Consortium out of the total exports was 100 percent in 1956, 98 percent in 1964, and 87 percent in 1974 (Table 8.2). During the period 1968-73 all the operators enjoyed a high rate of growth, but by 1974-75 exports slowed down because of the increase in oil prices and a worldwide economic recession.

The mechanism of crude oil exports from Iran is somewhat complex and the official NIOC Annual Reports do not really show the trend and importance of exports by various operators.

TABLE 8.2

Iranian Crude Exports by Various Operators, 1956-74
(thousands of tons)

Year	Consortium	Joint Ventures				NIOC	Total
		SIRIP	IPAC	LAPCO	IMINCO		
1956	24,494	--	--	--	--	--	24,494
1957	33,277	--	--	--	--	34	33,311
1958	37,774	--	--	--	--	48	37,822
1959	43,001	--	--	--	--	33	43,034
1960	48,680	--	--	--	--	45	48,725
1961	54,493	151	22	--	--	78	54,744
1962	61,402	312	13	--	--	102	61,829
1963	68,509	700	6	--	--	118	69,333
1964	78,134	1,255	77	--	--	105	79,580
1965	83,055	553	1,647	--	--	1,375	86,630
1966	95,329	818	2,187	--	--	1,300	99,634
1967	116,437	440	3,018	--	--	2,510	122,405
1968	127,203	767	3,717	330	--	2,060	134,077
1969	144,323	1,316	2,859	5,339	623	4,360	158,820
1970	163,797	1,572	1,845	6,666	2,786	5,880	182,546
1971	193,823	2,213	3,903	6,387	3,313	5,575	215,214
1972	215,219	883	3,772	5,709	3,711	11,835	241,129
1973	247,978	3,185	3,920	5,469	1,951	16,255	278,758
1974	247,775	3,302	4,001	5,666	1,732	20,945	283,421

Notes: (a) All figures are net crude exports; (b) NIOC figures for 1956-61 are for bunkering only. Thereafter, the export figures include sale of aviation fuel to Afghanistan. As of 1965, NIOC exports include the crude handed over by the Consortium as well as liftings from partnership ventures; (c) All Consortium data include crude delivered to the Abadan refinery for their own account.

Source: Compiled by the author from data in NIOC's Annual Reports 1967-73, NIOC's Marketing Department and the Statistical and Information Office of the affiliated companies.

(a) <u>Exports by the Consortium</u>. The production of crude by the Consortium is always higher than its exports of crude. This is so because some crude is exported as refined products from the Abadan refinery, and some delivered to the Abadan, Shiraz, and Tehran refineries for the account of NIOC for internal consumption. Also, under the 1967 Supplemental Agreement the Consortium has placed at the disposal of NIOC 20 million tons of crude over a period of five years, so that NIOC may export this crude to the markets where the Consortium participants are not operating. Table 8.3 provides detailed information on the crude produced, delivered, and exported by the Consortium.

(b) <u>Exports by Joint Venture partners</u>. According to its contract NIOC is able to ask its partners to deliver up to 50 percent of their production for its direct exports. If NIOC does not lift its share, the remainder will be sold in international markets at market prices or halfway prices, depending on the type of the contract. Unlike the Consortium, the foreign partners are mostly independent oil companies which sell their oil either on the spot market or to their long-term contractors. The only joint exporting activity is the supply of crude oil to the Madras refinery by IPAC (jointly owned by NIOC and AMOCO). The Indian refinery is owned in the ratio of 14:14:72 by NIOC, AMOCO, and the Indian government.

(c) <u>Exports by NIOC</u>. NIOC's crude exports have three components. First, NIOC exports directly very little of its own production--indeed, NIOC's main exporting activity in the past has been the supply of aviation fuel to Kabul and Kandhar airports in Afghanistan. The quantity of this export has been very small: 7,500 tons in 1969 and 11,200 tons in 1973 were exported by the Distribution Department of NIOC. A second source of crude for NIOC's exports has been its share of the partnership agreements. NIOC's liftings have frequently been less than its 50 percent entitlement. However, unlike the crude handed over by the Consortium, the partnership crude is not restricted to particular destination. NIOC can use this crude for political, economic, and strategic purposes to penetrate into international markets. Indeed, the partnership crude constituted the largest portion of NIOC's direct exports until 1973. In 1974 NIOC exported 1.1 million tons of crude to the Madras refinery and 1.9 million tons to Sasolberg refinery in South Africa.[4] The final source of crude for NIOC is the Consortium crude given under the 1967 agreement. Since the destination of the crude was restricted to Eastern Europe, NIOC was able to utilize only 9.2 million tons from the 20 million tons at its disposal. The opportunity to export oil to Eastern Europe arose because of the two-tier price system operated by the Soviet Union, whereby the price of oil sold to Eastern Europe is 50 to 80 percent above the price of Soviet crude sold to Western Europe.[5]

In October 1965 NIOC signed an agreement with Rumania under which
Iran was to export about $100 million worth of oil in ten years in ex-
change for machinery and goods required for Iran's development
plans. Since then further barter and cash agreements for the sale
of Iranian oil have been signed with other East European countries. The volume of exports to these countries was over 9.20 million tons in 1968-73 (see Table 8.3).

After the 1973 Sales and Purchase Agreement which replaced
the Consortium Agreement, NIOC's share of the Agreement Area was
no longer restricted to the East European market. As one can
see from Table 8.3, NIOC's liftings from the Consortium oilfields
increased from 2.1 million tons in 1972 to 12.6 million tons in 1974.

Three main points emerge from the analysis of NIOC's exports.
First, by 1974 NIOC had become a major force in the export of
Iranian oil. In this year NIOC exported 7.3 percent of the total exports and was the second largest exporter after the Consortium.
Value of the total exports by NIOC provided for per barrel income of
$1.45 in 1965 and $13.36 in 1974--twice as large as the Consortium
receipts in many years and up to four times larger than receipts
from partners. [6]

Secondly, NIOC's barter agreements with Eastern Europe represent an attempt at capturing a new and large market, which cannot
be fully supplied by the Soviet Union for much longer. It is impossible to estimate the dollar-per-barrel price of these exports, as
the Ministry of Economy does not release detailed figures of the
machinery imported. Moreover, the machinery specifications in
Eastern Europe differ from those in the West. However, it is fair
to speculate that NIOC prices have been much lower than those prevailing in international markets. NIOC is obviously suffering a loss
of revenue with regard to the opportunity cost of oil sold to Eastern
Europe. But this economic loss may be justifiable in terms of long-
term economic and political benefits to Iran.

Finally, NIOC's exports are mainly directed toward Europe
with the United States in second place. This reflects NIOC's prefer-
ence to take advantage of direct dealings with the European state oil
companies and the larger size of spot markets in Europe.

Whereas relatively good information is available on the destination of the Consortium's exports, little is known about the destination of partnership or NIOC exports, as the latter might change
hands a few times in the sea before reaching the final purchasers.

The Consortium's exports go to all the world continents. Japan
was the smallest importer of Iranian oil in 1962, but by 1971 she had
become the largest importer, having overtaken Western Europe in
1968. Other important changes in the pattern of demand include the
development of a sizable African market, which doubled its imports

TABLE 8.3

Components of NIOC's Direct Exports, 1965–74

(thousands of tons)

Year	Liftings from Joint Ventures[b]					Liftings from the Consortium	Total	Value of Exports	
	SIRIP	IPAC	LAPCO	IMINCO	Total			Millions of Dollars	Dollars per Barrel
1965	650	725	--	--	1,375	--	1,375	14.5	1.45
1966	380	920	--	--	1,300	--	1,300	13.0	1.37
1967	625	1,885	--	--	2,510	--	2,510	24.0	1.30
1968	145	1,585	--	--	1,730	330[a]	2,060	21.0	1.40
1969	--	2,225	585	--	2,810	1,550[a]	4,360	64.5	2.00
1970	--	2,825	560	--	3,385	2,495[a]	5,880	75.0	1.74
1971	570	2,380	475	--	3,425	2,150[a]	5,575	74.5	1.82
1972	3,495	3,145	2,705	400	9,745	2,090[a]	11,835	175.0	2.00
1973	505	2,545	3,815	1,430	8,295	{ 915[a] / 7,045 }	16,255	415.0	3.48
1974	410	2,630	4,060	1,200	8,300	12,645	20,945	2,050.0	13.36

[a]Represents the crude delivered by the Consortium under the 1967 agreement for exports to Eastern Europe.

[b]NIOC liftings from joint ventures may be added to the partners' exports in Table 8.2 to arrive at the total production of the joint ventures.

Source: NIOC's Marketing Department.

of crude between 1962-73. While North America was importing only
3.5 million tons of Iranian oil in 1970, its imports rose to 27.1 million tons in 1973, or 12 percent of total Iranian exports. This switch
by North America from North African oil to Middle East oil reflects
the attitude of Libya toward the oil companies and the increasing dependence of the United States on Middle East oil. The product markets are smaller, amounting to 15.1 million tons compared to 229.1
million tons of crude exports in 1973. In 1962 Asia (excluding Japan)
was the largest importer of refined products, but this lead was lost
to Japan and Africa in 1971.[7]

IRAN'S FUTURE EXPORTING PROSPECTS

The 1973 Sales and Purchase Agreement laid down specific
guidelines for Iran's exports from the southern oilfields for the duration of the contract 1973-93. These guidelines were based on the
size of Iranian proven and recoverable reserves as well as on the
size of the domestic market for oil products. Although it is doubtful
whether NIOC can follow the directions to the letter, nevertheless it
represents the Iranian recognition of two important facts. First, the
Iranian production has nearly reached its peak capacity and cannot
be substantially expanded and second, the prospects of new discovery
of oil bearing fields is small. Thus the largest group of oilfields
responsible for over 90 percent of Iranian production in the past two
decades will cease to export in 1993.

The total Khuzistan production is expected to reach its peak of
7.6 million barrels per day in 1978 and remain relatively steady for
six years. Thereafter the production will fall until 1993, when the
total production will only be capable of supplying the domestic market. NIOC's share of the exports will increase from 200,000 b/d in
1973 to 1.5 million b/d in 1981, holding steady for six years and then
dropping off. The share of the Consortium participants out of the
total exports will fall consistently as NIOC's liftings for both domestic sales and exports will rise.

It must be noted that the above figures are for the Southern oilfields and exclude the exports by two dozen or so foreign operators.
However, it is reasonable to speculate that these operators will, as
in the past, constitute only a small percentage of Iranian production,
providing for a total peak capacity of 8-8.5 million b/d.

The guidelines have, however, proved to be slightly exaggerated. The Iranian peak capacity is around 1 million b/d less than
the anticipated peak and the peak level cannot be sustained without
major gas-injection secondary recovery methods. The secondary
recovery method is expected to boost the recovery factor for oil in

place from 20-30 percent level attainable by primary depletion up
to 40 percent--with a corresponding increase in reserves. Initially,
seven of the larger oilfields (Agha Jari, Ahwaz-Asmari, Marum,
Gachsaran, Haft Kel, Bibi Hakimeh, and Paris) will carry out the
program. Haft Kel will receive the first injection of 300 million
cubic feet per day (cfd) from the gas cap of Naft Safid in 1975; other
fields will follow by 1977. By the end of the decade smaller fields
producing as little as 30,000 b/d may well be included in the program. Initially most of the fields are relying on associated gas, but
eventually supplies of natural gas will have to be developed for such
purposes. (For details, see Chapter 7.)

Thus valuable quantities of gas, which are flared in large volumes today and may well be exportable soon, must be used to sustain the productivity of the oilfields. To what extent this gas for oil
exchange is economically rational may be the subject of future research. Not only is increase in production on a substantial scale
technically hard to achieve, but, also, the slowdown in demand for oil
internationally has required a cutback in production of 10 percent in
the first quarter of 1975. Unofficial NIOC estimates put the most
likely output at 5.8 million b/d in 1975, 6.0 in 1976, and 6.2 million
b/d for 1977-82.

The high rate of growth of demand for oil products due to low
product prices and high rate of industrialization of Iran has been a
major factor in the decision to halt exports in the 1990s. As a matter of policy, Iranian reserves are not published by NIOC, though
such information is available in semiofficial publications and periodi-
cals, as well as in the text of speeches published in the national
press. One estimate in 1969 put the Iranian reserves at 45 billion
barrels,[8] while another estimate indicated a figure of 55 billion
barrels plus 14 billion barrels if secondary recovery methods are
employed.[9] But how flexible are the proven reserves of any country?
Proven reserves are governed by two factors: first, the finite quantity of oil in place and the recovery rate, which means that one day
this finite quantity will be entirely exhausted. Second, proven reserves are a function of the market price, and thus if the price of oil
rises, many oil wells which had previously been shut down as non-
commercial (because of high cost or small production) will become economically viable. Also, a great deal of offshore exploration and development will again become desirable. For example, in 1970 the
market price of Iranian crude was about $1.50 per barrel, with a 15
cent cost component or 10 percent of the price; but if the costs rose
to 50-70 cents, the marginal high cost production would decline. In
1975 the prices are set at $10.12 and a corresponding ratio would
make the cost of up to $1.00 per barrel acceptable (assuming the oil
companies are able to maintain their 50-100-cent profit margin).

Thus the Iranian proven reserves should rise because of the increase in market prices, but by how much no one knows yet. An estimation would require in part a study of all the wells closed down as uneconomic to see whether their cost pattern would allow their operation. A rule of thumb puts the Iranian reserves at present prices to about 100 billion barrels.[10] But in terms of relative size of reserves compared to other countries the Iranian position remains unchanged--Iran still has lower reserves than Saudi Arabia and possibly Iraq.

Let us see how important the 1973 Sales and Purchase Agreement is in the context of our upper and lower reserve boundaries. In the 1973-93 period Iran hopes to produce 42.5 billion barrels of oil for export and domestic consumption. If we take as the lower boundary 55 billion barrels and the higher boundary 100 billion barrels, at the projected level of domestic consumption of 0.5 billion barrels per year in 1993 will give the Iranian reserves between 25 to 100 years for the domestic consumption, with the true figure somewhere in between. If the market prices are increased or decreased our reserve values will change accordingly.

DOMESTIC SUPPLY OF OIL PRODUCTS

The crude oil used for supplying the domestic market comes from two sources: the Nafte-Shah oilfield, which is run by NIOC, and the crude delivered to the Shiraz, Abadan, Tehran, and Masjid-i-Soleiman refineries by the Consortium for the account of NIOC. The joint venture companies play no role in the domestic supply of crude oil and have no refining facilities in Iran.

At present there are four refineries and one topping plant in operation in Iran. These are Kermanshah, Shiraz, and Tehran refineries which are operated by NIOC for the domestic market, and Abadan export refinery and Masjid-i-Soleiman topping plant, which were operated by the Consortium until 1973. Although under the 1973 Sales and Purchase Agreement the management of Abadan was turned over to NIOC, for practical purposes the anticipated liftings of the former consortium from Abadan remain unchanged. Table 8.4 shows the operation of the Iranian refineries since 1956.

Abadan Refinery

The refinery is primarily export-oriented, but it also serves as a "balancing refinery" for Iran's domestic product requirements (see Chapter 11). Until 1973 it was operated by the Iranian Oil

Refining Company, a subsidiary of the Consortium, but under the operating agreements NIOC had priority for its domestic product needs.
The refinery is supplied with crude from the Consortium's Agreement
Area, which has an average gravity of almost 34 degrees of API. Gas
from the Marun area is piped to the refinery as fuel through a 16-inch
pipeline.

TABLE 8.4

Operation of the Iranian Oil Refineries
in Selected Years
(thousands of cubic meters)

Year	Abadan	MIS	Kermanshah[a]	Tehran	Alborz[b]	Shiraz	Total
1956	12,578	--	158	--	n.a.	--	12,578
1959	17,349	1,579	123	--	n.a.	--	19,051
1962	19,460	1,545	470	--	805	--	22,280
1965	20,312	2,009	523	--	21	--	22,865
1968	22,089	1,525	532	2,685	--	--	26,832
1971	23,194	1,830	701	4,939	--	--	30,664
1974	24,792	1,227	988	5,413	--	1,954	34,374

N.a. = not available.

[a]Figures for Kermanshah refinery include the operations of the
adjacent Nafte-Shah topping plant from 1961-73.

[b]Alborz topping plant ceased operations in 1965, Tehran and
Shiraz refineries started in 1968 and 1973 respectively.

Source: NIOC's Statistical and Information Office of the
affiliated companies.

Abadan refinery is a modern complex refinery, with a high degree of flexibility. It is one of the largest in the world and has a
capacity of 430,000 b/d. The refinery can produce 150,000 b/d of
middle distallates with a great flexibility between gas oil and kerosene,
ranging from 67 percent gas oil/33 percent kerosene to 40 percent gas
oil/60 percent kerosene. NIOC is entitled to lift 100,000 b/d of oil
products from Abadan, but these lifting ratios are subject to negotiations. The pricing of products for internal consumption is through
an arrangement between NIOC and the Consortium whereby NIOC pays

for the cost of crude and refining costs. In 1960 the cost of crude
was 25 cents and refining cost 66 cents. In 1973 the cost of crude
had declined to 10 cents and refining costs to 58 cents. The considerable reduction in unit costs was the result of increased oilfield
efficiency, economies of scale at the refinery, and the devaluation
of sterling in 1967 which lowered the 1 shilling per cubic meter producing and refining fees from 10.5 rials to 9.0 rials per cubic meter.

Under the 1954 Agreement the Consortium is obliged to export
at least 266,000 b/d from Abadan. In practice, however, these liftings have amounted to 300,000 b/d and the 1973 Agreement has left
these provisions unchanged.

One further point which deserves mention is that the imbalance
between the pattern of demand and supply in Iran has led NIOC to lift
high proportions of middle distillates from its entitlement at Abadan
to supplement the domestic supply. These products with relatively
higher world prices have caused NIOC a loss in terms of the opportunity cost of foregone export taxes (see Chapter 11).

MIS Topping Plant

Located at Masjid-i-Soleiman, this topping plant is 40 years
old. Until 1973 it was operated by the Consortium to supply fuel oil
to NIOC for the domestic market. Some fuel oil is returned to the
wells, while distillates are sent to the Abadan refinery. The fuel
oil for NIOC moves 98 kilometers through a 10-inch pipeline of
34,500 b/d capacity to the Ahwaz terminal. From there it is distributed by rail tank cars and trucks. Historically, the cost of fuel
oil to NIOC has declined from 31 cents a barrel in 1960 to 21 cents
a barrel in 1972. The decline in costs is primarily due to the sharp
reduction in crude costs which characterized the Consortium's operations in the past decade.

Kermanshah Refinery

The refinery was primarily constructed for the purpose of supplying the local market. The crude for the refinery is supplied from
the Naft-e-Shah oilfield on the Iran-Iraq border and flows through an
8-inch pipeline 260/km in length. Both the Naft-e-Shah oilfield and
the refinery were taken over by NIOC after the 1954 Consortium
Agreement, and until 1967 it was the only NIOC operated refinery.
The original capacity of the refinery was 5,000 b/d, but after extensive modernization the capacity was raised to 15,000 b/d in 1971.

Tehran Refinery

This was NIOC's first refining venture. The construction work
was completed in 1967 and the refinery came onstream early in 1968,
with the capital costs amounting to around $87 million. The crude
supply for the refinery comes from the southern oilfields through a
16.2-inch diameter Ahwaz-Tehran pipeline which is 750 km in length.
The refinery has a capacity of 85,000 b/d and its crude is either
Ahwaz crude or a mixture of 70,000 b/d of Ahwaz crude and 15,000
b/d of MIS-topped crude to produce the maximum quantity of middle
distillates. Ahwaz crude has 32° to 34° API gravity and produces a
relatively high percentage of middle distillates. Except for the first
two years, this refinery has always operated above its designed
capacity. It is planned to raise the capacity to 125,000 b/d by 1976/77.

Because of the increased demand for oil products in the capital,
a second Tehran refinery was designed and completed in March 1975.
The crude for this refinery is supplied through a second Tehran-
Ahwaz crude pipeline which was completed in late 1974 for the purpose of supplying future refineries in Tabriz and Isfahan as well as
the second Tehran refinery. This pipeline has diameters ranging
from 26 to 30 inches and is 735 km in length. The second refinery
has a capacity of 100,000 b/d and the two sister refineries will provide 225,000 b/d of products for the Tehran area.

The operation of the Tehran refinery has caused a decline in
fuel oil liftings of NIOC from the MIS topping plant. The refinery is
particularly important as it is located in the center of the largest
consuming area of the country.

Shiraz Refinery

Situated in the southern part of the country, the Shiraz refinery
commenced operation in November 1973. The refinery has a capacity
of 40,000 b/d and its crude input is supplied from the Gach Saran
field by a 10-inch, 230 km pipeline.

PLANNED REFINERIES

At present there are three refineries under construction or
study for domestic supply. Tabriz refinery with an output of 80,000
b/d, Isfahan refinery with an output of 200,000 b/d (with a design
similar to that of Tehran), and Neka refinery, near Meshed, with an
output of 130,000 b/d. All these refineries are expected to come onstream by 1978.

In the past two years a great deal of negotiations have been undertaken by NIOC and foreign oil companies for the construction of export refineries in Iran. Export refineries provide two main advantages for Iran: first, the economic benefit of exporting refined products by capturing the value added through refining, and second, the strategic advantage of controlling the crude as well as refineries in possible future maneuvers. Moreover, such ventures would open the way for going downstream at home rather than abroad. All three refineries originally planned for export with American, Japanese, and German concerns have now fallen through. The Irano-German refinery was to be located in Bushehr with a capacity of 500,000 b/d to export its products to Europe. But the European Common Market refused to grant the proposed refinery special import privileges and decided to subject it to import duty, which would make it uncompetitive with other refineries. The Iranians believe that the Bonn government did not do its best to obtain these privileges and were at the time of writing on the brink of canceling the agreement. The resistance of the industrialized countries to accept the movement of refineries to the oil-producing countries reflects their fear of an oil embargo, when the nonembargoed crude may not be refined for lack of refining capacity in Europe. Though Iran has never joined an embargo, she will find it hard to persuade the industrialized nations to build export refineries in Iran without airtight guarantees.

SUMMARY AND CONCLUSION

We have examined in this chapter the importance of Iranian oil production and export. It became clear that NIOC's own production has been negligible and its exports have mostly originated from the Consortium Agreement Area. NIOC's attempt to penetrate the international market has been commendable, but the progress was slow, as one might have expected. NIOC has been prepared to suffer losses of revenue by selling at lower prices, but this must be viewed as a market-penetration strategy, with possible long-term economic and political benefits for Iran.

With regard to the domestic supply of oil products, NIOC has been using its own crude from the Naft-e-Shah oilfield and the Consortium crude delivered to the various refineries for the account of NIOC. Unfortunately, because of the peculiar pattern of demand and the resulting imbalance between supply and demand, NIOC has been forced to lift more and more middle distillates from the Abadan refinery. These middle distillates have a higher export value than the other products and consequently NIOC has been foregoing potential tax revenues (see Chapter 11).

With the construction of the Tehran refinery, the Abadan refinery has become a balancing refinery, correcting the imbalance
between demand and supply. The Shiraz refinery, due for completion in 1971, was completed in November 1973. There are also
plans for three more refineries, in Tabriz, Esfahan, and Meshed.

NOTES

1. OPEC Statistical Bulletin and BP Statistical Review of the
World Oil Industry.

2. As described in Chapter 3, from July 1973, the Consortium's
activities were taken over by NIOC under the Sales and Purchase
Agreement. However, to maintain consistency the name Consortium
is retained to represent the production in the Khuzistan oilfields.

3. For details see Chapter 4 and the Oil Information Map of
Iran for locations.

4. The Madras refinery was operational in 1969 and has a
capacity of 50,000 b/d. NIOC and the American International Oil
Company are committed to supply 150-180 million barrels to this
refinery over a ten-year period. The Sasolberg refinery became
operational in 1971. It has a capacity of 50,000 b/d and is owned
by NIOC, TOTAL (French), and the South African Coal and Gas
Corporation.

5. Voprosy Economiki, April 1966, pp. 88-94, quoted in
G. W. Stocking, Middle East Oil (London: Penguin Books, 1971),
p. 195.

6. The term "value of exports" must be treated with caution.
It includes both the cash receipts for the sale of oil as well as Iran's
estimation of the worth of goods received through barter agreements.

7. Based on data from the Annual Report of the Iranian Oil
Operating Companies and NIOC's Statistical Office.

8. Reported in Iran Almanac and Book of Facts, Tehran, 1969.

9. Reported in International Petroleum Encyclopedia, 1970.

10. See the speech by Dr. R. Fallah in Kayhan International,
July 21, 1973.

9

SUPPLY II—MARKETING
AND DISTRIBUTION
OF OIL PRODUCTS

The purpose of this chapter is to trace the historical development of "Sazman-e-Pakhsh" or the Distribution Department of NIOC (DD). It will particularly concentrate on the post-1951 period, focusing attention on the problems faced by the Department. Moreover, this chapter will consider the constraints placed on the Distribution Department, either by the government, or by geographical and financial difficulties. This chapter will also attempt to indicate the changing structure of the DD during this 20-year period and the extent to which it has achieved its goals.[1]

HISTORICAL DEVELOPMENT

AIOC never considered the domestic Iranian market a very profitable one. It directed its efforts toward exporting the Iranian oil to Western Europe and to the large South Asian markets, where it could sell oil at prices agreed with the other majors. In all fairness one must admit that the lack of roads and railways restricted AIOC's chances of establishing a large distribution network in the country. This would have involved large infrastructure investments in roads and railways, which would not have been economically justifiable to a profit-making commercial entity. Thus the distribution network was small and barely met the Iranian government's minimum requirements.

After the nationalization of Iranian oil in 1951 and the subsequent establishment of NIOC, the Iranian government placed the expansion of the distribution network at the top of its list of priorities. It was the government's belief that the development of the Iranian economy would not be possible unless accompanied by the growing use of energy, and to achieve this, energy was to be made available to "anyone who needed it, anywhere in the country, and at reasonable prices."[2]

With the nationalization of Iranian oil, the Distribution Department was taken over by NIOC and was charged with the exclusive responsibility of oil product distribution in Iran. The DD's declared policy, with regard to oil product distribution, was as follows.

1. Oil products must be delivered and available to the marketing outlets.
2. The customer's requirements must be fully supplied.
3. The customer's requirements must be delivered when required.
4. Oil products must be of standard weight and quality when delivered.
5. Oil products must reach customers at standard prices, and the prices must be reasonable.
6. Oil product prices must be as cheap as possible to provide cheap energy for the expansion of all industries.[3]

IMPORTANCE OF THE DISTRIBUTION DEPARTMENT

The Distribution Department of NIOC is perhaps the most important organization within the Company. In 1973 the DD employed 7,958 people of the total 15,863 employed by NIOC--representing over 50 percent of the NIOC workforce. In terms of the total oil industry's labor force, the DD's share was 19 percent.

In 1973 the Distribution Department's sales in the domestic market exceeded $630 million, and its payments to the government in the form of sales tax and profit tax were larger than some of the partnership agreements, providing around 6 to 10 percent of the total oil revenues in the 1968-73 period (see Chapter 6).

The Distribution Department had to expand its distribution facilities within a short period to comply with the NIOC Constitution and government wishes. Because of this, it is the department most concerned with the overall planning of the economy, but its freedom to pursue autonomous distribution, sales, and pricing policies is severely restricted. It has to adapt its decisions to four factors: (a) physical possibilities, (b) financial possibilities, (c) NIOC's overall policy, and (d) the government's general planning strategy.

Because of the somewhat conflicting policies of NIOC and the "Plan Organization" (The organization responsible for the overall economic planning of Iran), the DD has had difficulties in complying with the wishes of both NIOC and the Plan Organization. Since

1969-70 the Plan Organization has moved toward exercising a tighter control over NIOC in general and the DD in particular. Because of this, there have been reports that the DD would become an autonomous organization by 1978, called the "National Iranian Oil Distribution Company," while remaining an affiliate of NIOC. This would allow the DD to work in close cooperation with the Plan Organization. Such close cooperation is clearly essential insofar as the construction of roads and railways is concerned. Also, to achieve the Plan Organization's policy of dispersing industries from Tehran and the main cities, and to encourage factories to move to underdeveloped regions, the cooperation of the DD is essential. One of the greatest problems facing the DD has been the position of the storage facilities in different areas. With the Plan Organization's cooperation, the DD will be able to assess or predict the population movements and the concentration of industries in particular areas, for which it could increase or decrease its storing capacity.

IMMEDIATE PROBLEMS FACING THE
DISTRIBUTION DEPARTMENT

Apart from the technical, managerial, and financial problems which the Distribution Department was facing, there were six important problems which the DD would have to solve to achieve its objectives:

1. The location of consumption centers. The northern and central areas of Iran have traditionally been the most prosperous regions with the largest consumption of oil products, while the less prosperous southern regions have been the producers of oil. In many cases the distance between the production and the consumption centers is over 1,000 kilometers (see Chapters 2 and 10). In 1968, 73 percent of the four main oil products (gasoline, kerosene, gas oil, and fuel oil) were consumed in the northern part of the country.[4] To transport oil from these areas to the centers of consumption, many new roads had to be built, railway lines laid, pipelines constructed, and road tankers purchased by NIOC. As we shall see in the course of this chapter, many of these problems have been solved in the span of 20 years, but some unpredictable factors, such as the closure of roads by avalanches (a common occurrence in northern areas) will still present difficulties in the future.

2. Lack of distribution centers. In 1950 there were only 691 points for the distribution of oil products in Iran, and these could not satisfy the domestic energy requirements of the country. It was the task of the Distribution Department to finance, organize, and control

the distribution centers. In 1973 the number of distribution points
had reached 12,051--a seventeenfold increase in 23 years (see
Table 9.2).

   3. Large fluctuations in seasonal consumption. Seasonal fluctuations in the consumption of oil products, particularly kerosene,
are so great as to cause a major storage problem. Kerosene, which
was previously used for lighting, is now used for cooking and heating,
and as such its distribution is essential for the well-being of the
people. Not only must adequate quantities of kerosene be reserved
for the cold season, but also the storage tankers must be located in
the centers of consumption themselves, since transport is always
difficult in the winter. A usual phenomenon in Iran with regard to
kerosene consumption is that in winter people not only buy kerosene
for their current consumption, but also hoard it in the expectation of
colder weather. The extent of the seasonal fluctuations in kerosene
consumption is shown in Table 9.1.

TABLE 9.1

Monthly Fluctuations of Oil Products Demand in 1967
(thousands of cubic meters)

	Kerosene	Gas Oil	Gasoline	Fuel Oil
January	228	158	72	177
February	205	142	56	167
March	169	161	70	164
April	143	166	67	165
May	111	189	71	184
June	97	187	73	164
July	93	209	80	181
August	94	209	80	179
September	104	196	76	195
October	133	203	77	213
November	174	184	73	213
December	257	179	59	236

Note: 1967 is chosen because it was a "typical year."

Source: Distribution Department Annual Report, 1967, p. 10.

We can see that while the three main products of gas oil, gasoline,
and fuel oil had a rather regular pattern of sales, the consumption
of kerosene has been erratic.

The Distribution Department is not reported to have even once
failed to deliver kerosene required by its customers. This has been
made possible by building large storage tanks in the main centers of
consumption. The provision of storage is costly, but this is a cost
which NIOC has been forced to accept under its Constitution. Unlike
the United Kingdom's National Coal Board, which sells coal more
cheaply in the summer than in the winter and thereby transfers some
of its storage costs to the coal users, NIOC is obliged to absorb the
full storage cost and is not permitted to pass any cost over to the
consumers. [5]

4. Importance of kerosene. Kerosene is by far the most important domestic fuel and this fact distinguishes the large Iranian
cities from those of Western Europe and America, where the major
domestic fuel is usually town/natural gas. Gas is distributed to
homes by means of pipelines, while the distribution of kerosene requires tankers and numerous sales outlets in the town, which naturally creates many more complex problems from the point of view of
planning.

5. Imbalance between supply and demand. The existing refinery pattern of production of the four main products is not in har-
mony with the demand pattern. That is, there is a shortage of refining capacity for some products, while there is an excess of other
products which cannot be exported due to lack of markets abroad.
(This will be discussed in detail in Chapter 11.)

6. Keeping up with the large annual increase in consumption.
The rate of expansion in the development of oil-based industries has
been very rapid indeed in the last decade. With an economic growth
rate of around 11 percent in the Fourth Plan period (1968-72) and a
rate of growth in consumption of oil products of 12.4 percent per
annum, keeping up with the supply and transport of oil products has
proved a difficult task for the Distribution Department. NIOC pre-
dicts a growth rate of 11 percent for the 1970-79 period, which
further complicates its problems. (Details in Chapter 10.)

ORGANIZATION OF DISTRIBUTION

In 1957 the DD decided to divide the country into seven geographical consumption centers--Tehran area, Khuzistan area, Esfahan area,
Khorasan area, Azarbayejan area, Kermanshah area, and Kerman
area. In 1966 the geographical division was changed, and the country

was divided into five areas. This continued until 1970, when it was
realized that there was too much centralized decision making, and
subsequently DD divided the country into nine areas. In 1973 the regional consumption of oil was as follows: Azarbayejan area (8.5 percent), Tehran area (35.5 percent), Khorasan area (8.1 percent),
Central area, comprising Shiraz and Esfahan area (12.7 percent),
Kerman and Zahedan area (4.6 percent), Kermanshah area (4.5 percent), Khuzistan area (8.8 percent), Rasht and Shahi area (8.9 percent), and Kashan-Ghom-Ghazvin-Hamadan area (8.4 percent). [6]

It must be emphasized that there has never been any particular
programming reason for the selection of the areas. DD has followed,
to some extent, the government's provincial divisions as well as its
own experience in allocating particular areas. [7] Since these divisions
do not follow the political divisions of Ostans (provinces) it is hard to
estimate the population densities accurately for any area.

Every area is a semiautonomous organization in terms of fore-
casting its requirements, but insofar as financial and pricing policies are concerned it follows NIOC's directives. In 1950 there were
691 distribution outlets, and in 1973 this figure reached 12,051. The
pace of expansion is shown in Table 9.2.

DEFINITION OF DISTRIBUTION CENTERS

Every oil area is divided into several districts. Every district has a number of branches, sales agencies, filling stations,
and rural/urban dealerships.

Branch

A branch or "bulk depot" is an integral part of the Distribution
Department. It is the largest distribution center and has large storage capacity. It is usual to find an NIOC run shop selling various
oil products alongside a branch, but the main purpose of a branch is
to supply large consumers and smaller distribution centers.

Sales Agency

Sales agencies are established in places where NIOC does not
find it profitable to open a branch. NIOC provides the assets (building, products, and other equipment) and the agent sells the oil products for NIOC at prices fixed by NIOC. The agent receives a commission proportional to the quantity of oil he sells. A sales agency

TABLE 9.2

Organization of the Distribution Centers in Selected Years

	1950	1960	1965	1966	1967	1968	1969	1970	1971	1973
Area	--	--	7	5	5	5	5	9	9	9
District	17	17	--	24	24	24	24	24	24	24
Branch	57	53	63	41	39	38	35	35	37	37
Sales agency	63	84	78	76	76	78	81	80	76	40
Filling stations*	147	172	238	251	266	323	340	365	390	418
	(--)	(--)	(48)	(68)	(80)	(134)	(155)	(185)	(215)	(275)
Rural dealers	215	927	3,614	5,048	6,179	6,960	7,543	8,233	8,066	9,226
Urban dealers	192	1,345	1,731	1,806	1,890	1,968	2,030	2,138	2,127	2,297
Total	691	2,589	5,731	7,251	8,481	9,396	10,058	10,884	11,469	12,051

*Figures in brackets represent privately owned filling stations.

Source: For 1950, 1960, 1965, 1967, the Distribution Department Progress Report 1968. Figures for 1969–73 are based on data which have been collected from the Annual Reports and brought to a common base.

is usually quite small and does not have large storage capacity. The agent must have a good local reputation, and he must present cash or property as surety against any damage or loss to NIOC properties.

Service Stations

Until the early 1960s, all the filling stations belonged to NIOC. In 1965 there were 48 privately run filling stations, compared with 185 NIOC filling stations. In 1967, the Board of Directors of NIOC decided to halt any increase in the NIOC-run filling stations and encourage private enterprise to supply the required number of filling stations. This was intended to have two important results: first, it was to have constituted a step toward a greater integration of the economy with the oil industry, and second, it was to bring about the establishment of filling stations on a more economical basis. In 1973 there were 275 privately owned and 143 NIOC-owned filling stations. The procedure for allocating a license to a private filling station is as follows: first the Distribution Department decides on a particular area and then on a particular road. Then the DD advertises in the national press the fact that anyone who owns a piece of land in accordance with the required specifications, and can present reputable guarantors, can apply for a license. After one person or group of persons are chosen, NIOC builds the filling station and supplies all the machinery and pumps. The filling station will then buy oil from the DD and sell it at fixed prices. No competition is allowed between the filling stations, although there is a great deal of nonprice competition. It is interesting to note that unlike European countries, the private filling station does not receive a discount of prices from NIOC. It receives a commission on the quantity of products it sells. The nature of the commission structure of the filling stations is highly unusual compared with the Western conception of "quantity discount," as will be seen later.

Rural/Urban Dealers

These are private retailers who buy oil from the DD at special discounts for resale at fixed prices and receive a commission depending on their sales volume. Unlike sales agencies, they do not represent NIOC, have their own installation, and maintain no direct links with the DD. These dealers represent the largest increase of distribution centers in the past 20 years. An important component of rural dealers are the cooperatives, which were set up after the implementation of the Iranian land reforms. Rural cooperatives were

negligible until 1965. From 1965 to 1970 their numbers increased
from 2,119 to 5,590, constituting the largest growth in the dealers.

SALE OF OIL PRODUCTS THROUGH VARIOUS
DISTRIBUTION CENTERS

Table 9.3 shows us the channels through which the four main
products were sold in 1972.

TABLE 9.3

Sale of the Four Main Products Through Various
Distribution Centers in 1972
(percents)

Distribution Center	Kerosene	Gasoline	Gas Oil	Fuel Oil
Dealers (urban/rural)	77.6	6.3	16.2	8.9
Cooperatives	8.8	0.1	4.3	0.5
Private filling stations	2.4	28.2	22.1	0.5
Subtotal (private enterprise)	88.8	44.6	42.6	9.9
NIOC filling stations	1.9	41.2	10.7	0.1
Direct from depot (including agencies)	9.3	14.2	46.7	90.0
Total	100.0	100.0	100.0	100.0

Source: Distribution Department's Statistical Office.

Considering each product in turn, we can see that

(a) Kerosene sales were largely through private enterprise,
     with retail outlets constituting the principal sales outlet.
     Filling stations played a minor role in the sale of kerosene.
(b) Gasoline sales are equally divided between private enter-
     prise and NIOC's own outlets. The largest quantity of gaso-
     line it sold has been through the filling stations. Dealers
     and cooperatives have played a minor role in the sales of
     gasoline. [8]

(c) Gas oil sales are dominated by direct sales from depots,
     while filling stations play a secondary role. Dealers pro-
     vide 16 percent of the total gas oil sold in Iran.
(d) Fuel oil sales are made almost exclusively from NIOC
     branches or bulk depots; private enterprise dealers have
     less than one-tenth of the market.

COMMISSIONS

The rate of growth in rial payments for commissions between
1962 and 1969 outstripped those of any other single cost element. In
1968 and 1969, commission payments accounted for a quarter of the
total costs of the DD. Table 9.4 shows the commission payments in
rials per cubic meter.

TABLE 9.4

Commissions in Selected Years
(rials per cubic meter)

Product	1962	1966	1967	1968	1969
Kerosene	143.5	137.5	144.2	145.5	144.1
Gas oil	16.5	52.7	60.4	64.4	70.8
Gasoline	24.9	52.1	60.9	70.0	79.7
Fuel oil	4.2	10.9	11.0	11.7	13.1
Iranol	54.8	80.4	106.8	91.8	136.9
Bitumen	1.2	1.4	1.4	1.4	1.4
All other	4.1	47.9	1.3	1.1	3.4
Average	47.0	59.4	62.1	64.2	68.3

Note: Figures are based on the total sales volume.

Source: Calculated from the data supplied by NIOC.

Clearly, the commission payments have increased considerably
over the seven-year period. It is hard to explain why this has happened, particularly because the DD does not differentiate between the
commission received by various outlets and its data are for total
commission payments and volumes. One reason may well be the

expansion of private filling stations which supply a large quantity of
gasoline and gas oil (see Table 9.5). Another reason for the behavior
of the commission payments is the rather peculiar structure of commission rates in Iran.

TABLE 9.5

Basic Commission Structure (1969-72 Average)
(rials per cubic meter)

Product	Private Filling Stations	Urban/Rural Dealers Cooperatives and Others	Agencies
Gasoline	150*	100	100
Kerosene	150	150	80
Gas oil	160*	100	70
Fuel oil	--	100	60

*Averages based on 1969-72 sample survey.

Source: Statistical Office of the Distribution Department.

There are basically three kinds of commission recipients in
Iran--agencies, dealers, and private filling stations. The commission structure for each group is shown in Table 9.5. Unlike the
common practice in Western Europe and the United States, dealers
and agencies receive no quantity discounts. The quantities sold do
not always have a bearing on the commission they receive. But what
is even more surprising is the commission rate paid to private filling stations. Private filling stations receive a flat commission rate
of 150 rials per cubic meter sold. But the commissions received
for the sale of gasoline and gas oil are inversely related to the sales
volume, as shown in Table 9.6.
    This arrangement is very unusual in any business practice.
The DD claims that this kind of arrangement will stop any private
filling stations from capturing the customers of the neighboring
NIOC/private filling station. In general the author was not satisfied
by the explanations given by the various NIOC officials in this
matter.

TABLE 9.6

Private Service Stations' Commissions
for Gasoline and Gas Oil

Monthly Sales, Cubic Meters	Rate, Rials per Cubic Meter
Up to 50	300
50-400	130
400-1,500	100
Above 1,500	80

Source: Statistical Office of the Distribution Department.

IRANIAN OIL TRANSPORT NETWORK

The Iranian oil transport network comprises pipelines, railways, road tankers, and barges.

Pipelines

The Abadan-Ahwaz product pipeline moves products from the Abadan refinery to the Ahwaz terminal for local distribution, and also supplies the product pipeline to the north. The 12-inch line extends over 128 kilometers and was put into operation in 1958. It has one pumping station of 3,000 horsepower located in Abadan. The line had a maximum capacity of 67,000 b/d in 1971. A new 8-inch pipeline, which paralleled the existing line, is currently under construction and is scheduled for completion in the early 1970s. Without the addition of new pumping facilities, the combined capacity of both lines will be 89,000 b/d. This capacity could be increased to 130,000 b/d by using an additional pumping plant at Abadan and a new pumping station between Abadan and Ahwaz. The cost of these pumping facilities is estimated at 426 million rials (U.S. equivalent, $5.6 million).

The 821-km, 10-inch Ahwaz-Rey line is the key link with the main consumption center of Tehran, and also serves distribution depots along the route as well as the branch line to Esfahan. It started operations in 1957 with five pumping stations, having a total of 8,160 installed horsepower. In 1961 an expansion program was completed which added nine more pumping stations with 14,130 total

TABLE 9.7

NIOC Product Pipelines--Technical, Operational, and Capital Investment Data

Product Pipelines	Diameter, inches	Length, kilometers	Maximum Barrels per Stream Day	Present Capacity Average[a]	
				Thousands of Barrels/ Calendar Day	Thousands of Cubic Meters per Year
Existing Lines					
Abadan-Ahwaz	12	128	67.0	66.2	3,842
Ahwaz-Azna	10	482	54.0	53.0	3,076
Azna-Esfahan	6	234	11.7	10.5	609
Azna-Arak	10	87	50.0	48.6	2,820
Arak-Rey	10	252	44.0	39.5	2,292
Rey-Gazvin	8	150	18.8	16.9	981
Gazvin-Kasht	6	175	12.0	10.8	627
Rey-Shahrud	8	375	12.6	11.3	656
Shahrud-Mashhad	8	444	11.7	10.5	609
Pipelines under construction					
Abadan-Ahwaz (in operation 1971)	8	128	22.0	19.8	1,149

Product Pipelines	Construction Costs			Average Estimate		Year in Operation	Progress Status	Cost of Expansion	
	Year of Commission	Millions of Rials	Millions of Dollars	Thousands of Barrels per Calendar Day	Thousands of Cubic Meters per Year			Millions of Rials	Millions of Dollars
Existing Lines									
Abadan–Ahwaz	1958	432	5.7	--	--	b	b	--	--
Ahwaz–Azna	--			58.0	3,366	1973	c	--	--
Azna–Esfahan	1958	388	5.1	15.0	870	1971	c	160	2.1
Azna–Arak	1957	3,832	50.9	53.4	3,099	1973	c	--	--
Arak–Rey	1960			43.4	2,519	1973	c	904	11.9
Rey–Gazvin		678	9.0	--	--	b	c	--	--
Gazvin–Kasht	--			13.6	789	1971	c	152	2.0
Rey–Shahrud				14.5	841	1971	c	129	1.7
Shahrud–Mashhad	1961–62	1,428	18.9	--	--	--	Plans		
Pipelines under construction						--	--	--	--
Abadan–Ahwaz (in operation 1971)	--	--	--	--	--	1971	c		

[a] Assuming 90 percent load factor.

[b] No expansion plans.

[c] Under construction.

Note: A new Ahwaz–Tehran pipeline was constructed in 1974 to feed the second Tehran refinery, Tabriz refinery, and other projected refineries. The pipeline is 735 km long and includes 228 km of 30-inch diameter pipe and 507 km of 26-inch pipe.

Source: National Iranian Oil Company

installed horsepower. Three pumping stations are electric and nine
are diesel driven. The present maximum capacities of the various
sections are shown in Table 9.7. A program of electrifying all the
pumping stations is under way, and this is expected to increase the
pipeline capacities by 10 percent.

The Azna-Esfahan line has a diameter of 6 inches and moves
products over a distance of 234 kilometers. It was put into operation in 1958 with one pumping station of 1,080 horsepower, and has
a present maximum operating capacity of 11,700 b/d.

The 325-kilometer Rey-Rasht pipeline was operational in 1960,
with one electric pumping station of 900 horsepower, located in Rey.
The products are received from Rey (on the outskirts of Tehran) and
transported for distribution to Gazvin and Rasht. The maximum
capacity of the Rey-Gazvin section is 18,800 b/d (8-inch line),
whereas the 6-inch diameter section from Gazvin to Rasht has a
capacity of only 12,200 b/d. A new pumping station with 700 installed horsepower is under construction at Gazvin, and will cost an
estimated 152 million rials (U.S. equivalent $2.0 million). When in
operation, it will increase the capacity of the 6-inch line to Rasht
to 13,600 b/d.

The 8-inch Rey-Meshed product line, with a length of 819 km,
has been operational since 1962 with two pumping stations having
2,280 total installed horsepower. The maximum operating capacity
of the first section from Rey to Shahrud is 12,600 b/d, which will
be increased to 16,200 b/d when a new 1,400-horsepower pumping
station at Semnan, costing an estimated 129 million rials (U.S.
equivalent $1.7 million), is completed. The second section from
Shahrud to Meshed has a capacity of 11,700 b/d. The throughput of
the pipeline can be further increased by additional pumping stations.

Road Tankers

This mode of transport is an important link in the distribution
system. Road tankers serve local retail outlets and wholesale consumers and also move products, including fuel oil, long distances
in areas not served by pipelines or railways. Due to their mobility
and speed they are also used as emergency transport means if shortages develop because of bottlenecks or breakdowns in pipeline or
rail transport. A large portion of the total road tanker fleet is
owned by private enterprise. In 1960 there were 355 NIOC-owned
tankers, compared to 904 tankers under contract. In 1969, the number of NIOC-owned tankers was slightly reduced to 340, while the
number of privately owned tankers had risen over 2.5 times to
2,450. The growth in privately owned tankers is clearly due to a

deliberate NIOC policy of encouraging private enterprise to take over
more responsibility for distribution.

Railways

In 1970 the Iranian State Railway operated a network of 3,700
kilometers. The main lines form a "T" with Tehran at the intersec-
tion. The most frequent connection is between Khoramshar and
Bandar Shahpur, on the Persian Gulf, via Tehran to Bandar Shah on
the Caspian Sea, over a distance of 1,440 kilometers. It is on this
line that the railway moves (on behalf of NIOC) a considerable amount
of fuel oil from the loading point in Ahwaz to the north. Another
major line connects Tehran with Tabriz (736 kilometers) and from
there to Turkey over a distance of 224 kilometers. From Tabriz a
railway connection exists to Julfa on the Soviet border. The third
section of the "T" leads from Tehran to Meshed over a distance of
926 kilometers. A plan exists to connect Qom with Zahedan to form
a link with the Pakistani Railway system. In the early 1950s and
early 1960s, the Iranian Railways played an important role in the
Iranian oil transport network. This role, however, has now been
greatly reduced because they do not operate efficiently and reliably.
The utilization ratio of existing traction equipment is extremely low
in comparison with other countries and the maintenance of fixed installations is not properly carried out. [9] As a result, rail transport
is slow, delays are frequent, and NIOC has no direct control over
the situation. However, as long as there is a major fuel shortage in
the north which must be covered from the MIS topping plant or the
Abadan refinery and which cannot be moved by pipelines, the railways
are necessary and could contribute much to an efficient and low-cost
product distribution in Iran. In addition, the railway tank cars have
a much longer life than road tankers, are cheaper to run, and carry
a much larger capacity--namely an average of 43 cubic meters, compared to 16 cubic meters for road tankers. There are 1,060 railway
tank cars in Iran. Of these, about 10 percent are usually out of operation due to maintenance and repairs. Another 10-15 percent are
allocated to the military or for the private use of the railways. This
leaves NIOC with 800 to 850 cars under normal conditions, which is
equal to a capacity of 34,400 to 36,500 cubic meters. This may be
compared with a road-tanker capacity in 1969 of about 43,000 cubic
meters, one-eighth of which is owned by NIOC.

Barges

A few chartered barges supply petroleum products in bulk and
in drums to towns along the Persian Gulf, the most important of which

is Bandar Abbas. These barges, with a capacity of 580 to 1,740
cubic meters, are moving a small but increasing amount of oil
products out of Abadan. The distance from Abadan to Bandar Abbas
and Bushehr is 100 kilometers and 400 kilometers respectively.

Relative Importance of Various Means of Transport

The relative contribution of each means of transport is shown
in Table 9.8. It will be seen that pipelines constitute the most important carrier of oil products in Iran. After pipelines, privately
owned tankers and the railways are in the second and third places.
NIOC has cut down on the operation of its own road tankers; in 1971
it closed down the Kermanshah Tanker Group serving the western
and central districts of the country and transferred its responsibili-
ties to contracted road tankers. The impact of this closure can
clearly be seen in Table 9.8. NIOC gave up its operation of barges
in 1963 and transferred this activity to private-enterprise shipping.
The use made of the transport system increased from 1,824
to 12,469 million ton-kilometers between 1957-73--a sevenfold increase in 16 years. Since the early 1960s pipelines carried the
largest volume of oil. This period is marked by a large decline in
the relative importance of railways from 43 percent in 1957 to 13
percent in 1973, and a very rapid rise in the use of pipelines from
27 to 58 percent in the same period.[10] The share of privately owned
road tankers remained relatively steady despite annual fluctuations,
while NIOC road tankers became the least important. Barges have
remained a minor factor throughout the period.
The rate of growth of pipeline use was 29 percent in 1967 and
over 7 percent in 1973. The drop in the growth rate of pipelines can
be explained by the construction of the Tehran refinery in 1967.
There was a smaller need for building new pipelines or using the
existing lines to the maximum capacity, as Tehran--the largest consuming center--and nearby areas were supplied by the local refinery. Privately owned road tankers maintained a high rate of growth
throughout the period. In total, the various means of transport
showed a growth rate of 14 percent in 1967, 9 percent in 1971, and
22.5 percent in 1973.
In general, one can conclude that the oil transport network in
the 1957-73 period was dominated by increased pipeline use and decreased railway use. But was there any economic advantage in the
increasing use of pipelines over other means of transport? To examine this we have to look at the cost structure of the transport
network.

TABLE 9.8

Usage and Relative Importance of Various Means of Transport for Selected Years
(million ton-kilometers and percents)

	1957		1967		1971		1973	
	Million Ton- Kilometers	Percent	Million Ton- Kilometers	Percent	Million Ton- Kilometers	Percent	Million Ton- Kilometers	Percent
Pipelines	479	27	4,037	67	6,773	72	7,284	58
Railways	779	43	943	15	829	9	1,615	13
Road tankers (private)	400	22	883	15	1,565	17	3,028	24
Road tankers (NIOC)	102	6	150	2	93	1	93	1
Barges (private)	--	--	49	1	144	1	449	4
Barges (NIOC)	64	2	--	--	--	--	--	--
Total	1,824	100	6,062	100	9,404	100	12,469	100

Source: Based on data from the Distribution Department's Annual Reports 1966-73.

THE STRUCTURE OF TRANSPORT COSTS

Transport costs constitute a major component of the total cost of oil products. Indeed, the relative importance of transport costs has remained relatively steady in the past decade. As we can see from Table 9.9, transport costs have constituted between 42 and 43 percent of total costs in 1962 and 1972. Furthermore, since the figures for transport cost of crude are excluded from the data, one can speculate that the real contribution of transport costs to total costs is well over 50 percent. In light of this it is essential to consider the variation in the costs of each means of transport so that the economic rationality of the usage of various modes of transport may be evaluated. Table 9.10 shows the average costs of each mode of transport in rials per ton-kilometer for 1957-73.

TABLE 9.9

Costs in the Iranian Oil Industry, 1962 and 1972

	1962		1972	
	Millions of Rials	Rials per Cubic Meter	Millions of Rials	Rials per Cubic Meter
Cost of product[a]	1,972	443	4,097	321
Container cost	312	70	579	47
Transport cost of products	2,667	599	5,928	465
Distribution costs[b]	1,269	285	3,635	285
Total costs	6,220	1,397	14,239	1,118

[a]Includes the cost of crude, transporting crude to refineries, and the cost of refining.

[b]Includes wages and salaries, commissions, and depreciation.

Source: The Statistical Office of the Distribution Department.

TABLE 9.10

Average Cost of Various Means of Transport in Selected Years
(rials per ton-kilometer)

	1957	1967	1971	1973
Pipelines	0.99	0.35	0.23	0.20
Railways	1.30	1.12	1.15	1.12
Private road tankers and barges	1.87	1.47	1.30	1.25
NIOC road tankers	2.36	1.70	3.20*	3.00*
Average cost	1.37	0.67	0.56	0.53

*The large increase was due to the closing down of the
Kermanshah Tanker Group owned by NIOC.

Source: Statistical Office of the Distribution Department.

It can be seen that the average costs of all means of transport
have declined. As expected, the largest decline in costs is attributed
to the pipelines which show a drop of about three quarters in costs in
16 years. The largest drop in the pipeline cost was in 1958, when
costs equaled half the 1957 costs. This drop was due in large part
to the opening of the first trans-Iranian pipeline for operation. The
pipeline cost then remained relatively steady until 1967, when the
second trans-Iranian pipeline--that supplying the Tehran refinery's
requirements--was ready for operation. It led to a fall in the pipeline costs to an average of 0.35 rials in 1967, and finally to 0.20
rials in 1973. It is important to note that even in 1957, when heavy
investment in pipeline construction was taking place, the average
pipeline cost was the lowest of all alternatives, indicating that the
expansion of the pipeline network was economically sound. The average railway cost also declined, but the decline was only 14 percent
in 16 years. The first drop came in 1959, but thereafter the cost
was steady until 1964. In 1964 the cost declined to 1.08 rials and
stayed constant for three years. As the railway freight charges increased, the average cost rose to 1.12 rials in 1967, and remained
relatively steady until 1973. The cost of privately owned road transport, tankers, and barges has steadily dropped over this period, and
considering that they were the second largest carriers in the 1957-
73 period, their operations seem to have been satisfactory. It is also

an indication of the lack of monopolistic power in the private road
transport and barge industries, and shows good judgment on the part
of NIOC to leave this section of the transport system in the hands of
private enterprise. NIOC's road tanker costs declined steadily, but
this decline was not significant, particularly in view of the small
amount of oil products they carried. In total, average costs dropped
by 61 percent in this period.

Pipeline costs show a continued drop of 12 percent in 1967 and
13 percent in 1973, which is by all standards quite significant. Railways show an increase of 4 percent in costs in 1967, but there was a
decline in costs in 1973. The cost performance of NIOC-owned road
tankers is disappointing, and justifies the closing down of the Kermanshah Tanker Group in 1971 (see Figure 9.1).

SUMMARY AND CONCLUSION

We have seen in this chapter the difficulties and constraints
faced by the Distribution Department of NIOC during the period after
nationalization. Because of its importance, the DD was subject to
constraints that were not only geographical, technical, and financial,
but also political.

Given the difficulties it had to face, the expansion of the distribution centers and transport facilities must appear as a great success story. The DD expanded its distribution outlets seventeenfold
in 23 years and provided oil products of standard quality at fixed
prices. It managed to bring oil products to any customer who required them, any time, anywhere in the country. It bore the burden
of high storage tanks for kerosene so that no one would be left without it.

So far as transport policy was concerned, it was economically
sound to invest in pipelines, with consequent reduction in pipeline
costs. The overall success of the transport policy was not carried
over into the area of the operation of the NIOC-owned road tankers
and barges. The barges were abandoned in 1963, and road tankers
are currently being phased out. NIOC correctly decided that it would
be a sound policy to leave these parts of the transport network in the
hands of private enterprise. The soundness of this policy can be
judged from the points listed below.

1. During this period, not only were the costs of the privately
   owned tanks and barges lower, but also there were consis-
   tent reductions in average costs.
2. It brought about a closer integration between the national
   economy and the oil industry.

FIGURE 9.1

Comparison of Various Transport Costs
1957 = 100

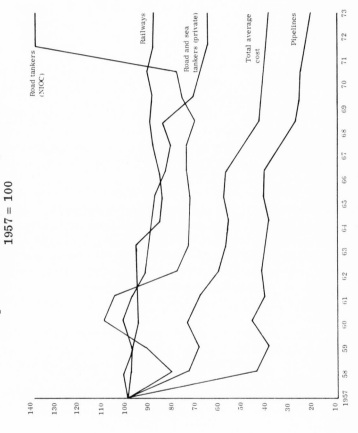

Source: Statistical Office of the Distribution Department.

3. It decreased the burden of administrative and centralized decision making of NIOC.

In general, it is wrong to appraise NIOC's activities on a purely commercial basis. It has to be understood that as a part of a public corporation the DD has different objectives from those of a commercial entity, and therefore needs to be judged in a different framework. Given the constraints referred to above, it must be said that the DD has been successful in achieving its goals.

NOTES

1. Most of the data presented in this chapter have been collected from the statistical office of DD. DD is unable or unwilling to present data for the 1951-57 period, when the operation of the Department was disrupted by the nationalization and its aftermath.

2. NIOC's Constitution (Policy Section).

3. Policy declaration of the Distribution Department, quoted in F. Saadat, An Economic Geography of Iranian Oil, Vol. II (Tehran: 1968). In Persian.

4. H. Farkhan, "A Forecast of the Sale of Oil Products in the Next Decade," in Seyri Dar Sanaat Naft-e Iran (A Journey through the Iranian Oil Industry), NIOC Public Relations Office, 1970, pp. 130-61.

5. The Distribution Department currently holds, on the average, about 37 days supply of oil products based on the last year's consumption. For kerosene, 52 days supply is held.

6. For details see Saadat, op. cit., pp. 403-09, and DD's Annual Reports.

7. In the next few years DD is expected to adjust the areas by using programming methods.

8. Urban and rural retailers sell gasoline in 5-liter containers at higher prices.

9. Quoted from a government report, "A Survey of the Iranian Economy," published by Iran Trade and Industry, Tehran, 1969, p. 78.

10. The year 1973 was a particularly difficult year for DD. Because of a very cold winter, product demand grew by an unprecedented 18 percent. As pipeline capacity could not cope with this demand, increasing use was made of railways, road tankers, and barges. However, railway use is not expected to exceed its 1971 ratio in future years.

10

DEMAND FOR
OIL PRODUCTS

Chapters 8 and 9 examined the sources of supply and the distribution of oil products in the domestic market. This chapter is concerned with the demand conditions in the country during 1951-74. The introduction to the chapter discusses the pattern of energy consumption. This is followed by a historical analysis of demand and the factors affecting it, the development of the oil markets, and the role of oil in household expenditure.

DEMAND FOR ENERGY AND THE ROLE OF
PETROLEUM PRODUCTS

In the Western world the primary source of energy for fueling the "Industrial Revolution" was coal. Early in the twentieth century the attention of the industrialized countries was turned toward oil and its derivatives following the discovery of numerous commercial and industrial uses of petroleum products and hence their availability as an alternative source of energy. Progress in refining and the availability of oil in large quantities facilitated a switch from coal to oil. Today in most industrialized countries of the world petroleum products account for at least 45 percent of their energy requirements. Table 10.1 shows the pattern of energy consumption in the industrial countries of the world.

Generally, in the developed world, solid fuel has given way to petroleum products. This transformation is particularly striking in the case of Japan. However, within this general framework, several points need to be emphasized.

TABLE 10.1

Pattern of Energy Consumption in the Industrial Countries
(percents)

Area	Year	Solid Fuel	Oil	Natural Gas	Hydroelectricity	Nuclear	Total
United States	1955	31.4	40.1	27.3	1.2	--	100
	1965	23.7	40.4	34.5	1.4	--	100
	1970	21.1	44.6	32.8	1.4	0.1	100
	1973	17.8	47.2	32.8	1.4	0.8	100
Western Europe	1955	78.2	18.1	1.0	2.7	--	100
	1965	49.3	44.3	2.7	3.5	0.2	100
	1970	29.4	61.1	6.1	3.0	0.4	100
	1973	22.1	63.9	10.8	2.7	0.5	100
Japan	1955	70.0	20.6	0.3	9.1	--	100
	1965	36.9	56.5	1.5	5.1	--	100
	1970	22.2	74.1	1.0	2.6	0.1	100
	1973	15.4	80.4	1.6	2.2	0.4	100

Sources: S. H. Schurr and P. T. Homan, Middle Eastern Oil and the Western World Prospects and Problems (New York: Elsevier, 1971), statistical appendix, pp. 161-77 for 1955 and 1965 figures. For 1970 figures see B.P. Statistical Review of the World Oil Industry, London, 1970 and 1973.

1. The process of replacing coal by oil will not continue indefinitely. At some point in time, the share of solid fuel will
   become stabilized because of the security of supply, and for
   social and employment reasons.
2. Natural gas, which has been playing an important role in
   the American energy consumption, is expected to become
   increasingly used in Western Europe.
3. Although hydroelectricity and nuclear power have been
   relatively unimportant in the past, ultimately the latter is
   expected to become the substitute for oil.

ROLE OF PETROLEUM PRODUCTS IN
ENERGY CONSUMPTION OF IRAN

As one might expect, the demand for energy in Iran is dominated by petroleum products and natural gas. Hydroelectricity,
coal, charcoal, wood, and animal matter play subordinate roles.
Table 10.2 shows the pattern of energy demand in Iran for selected
years. Certain observations can be made from the table.

1. Petroleum products have increased their share from 63
   percent in 1960 to 68 percent in 1972, but it is expected
   that larger utilization of natural gas and hydroelectricity
   will reduce the share of petroleum products to 60 percent
   by 1982.
2. A striking feature of the table is the large share of the
   natural gas consumption in Iran. While the industrial and
   residential uses of natural gas are very small indeed, the
   major portion of natural gas is used for oil-production purposes and as a refinery fuel. Although the share of natural
   gas has declined throughout this period, it is expected that
   its use will reach 34 percent of the market in 1982, when
   the plans for laying pipelines in the large cities are due to
   be completed.
3. Coal, wood, and animal matter have declined in importance,
   and this trend is expected to continue.

A word of caution is in order regarding the reliability of the
data in Table 10.2. While it is easy enough to estimate the demand
for petroleum products, natural gas, hydroelectricity, and coal because the sources of supply are few and the data well recorded, it is
much more difficult to estimate the consumption of wood or the use
of animal matter. This is due partly to the multiplicity of suppliers,
and partly to the fact that as these products are mainly used in the

TABLE 10.2

Pattern of Energy Demand in Iran, 1960–82

	1960	1963	1965	1969	1972*	1977*	1982*
Petroleum products	132.0	161.0	200.0	328.5	449.0	674.0	963.5
	(63)	(67)	(71)	(74)	(68)	(62)	(60)
Natural gas	39	40	44	60	142	300	590
	(18.0)	(16.5)	(15.0)	(14.0)	(22.0)	(28.0)	(34.0)
Wood and charcoal	24.5	20.0	19.0	17.0	15.0	13.0	12.0
	(11.0)	(8.5)	(7.0)	(4.0)	(3.0)	(1.0)	(1.0)
Animal matter	11.5	10.0	9.0	7.0	6.0	4.5	3.5
	(5.0)	(4.0)	(3.0)	(1.0)	(1.0)	(1.0)	--
Coal	6.0	5.5	6.5	22.0	7.0	16.0	15.0
	(3.0)	(2.5)	(2.0)	(2.0)	(1.0)	(1.0)	(1.0)
Hydroelectricity	0	3.5	5.5	22.0	35.0	75.0	75.0
	0	(1.5)	(2.0)	(5.0)	(5.0)	(7.0)	(4.0)
Total	213.0	240.0	284.0	441.5	654.0	1,082.5	1,659.0
	(100)	(100)	(100)	(100)	(100)	(100)	(100)

Note: First number indicates figures in trillions of BTUs, with second number in parentheses indicating percents.

*Estimated.

Source: Stanford Research Institute's Report to the Government of Iran, Long Term Energy Policy in Iran, 2, 1970 (unpublished).

villages, adequate records are not kept. The Stanford Research Institute claims that these data were collected by the use of sample
surveys, but provides no indication of the methods used in estimating
these data.

The four main products--gasoline, kerosene, gas oil, and fuel
oil--account for the major portion of oil products used in Iran. In
1957 these products accounted for 96 percent of the total oil product
demand of the country. This ratio fell to 91 percent in 1961 and has
not altered since. This indicates the significance of these products
in the pattern of energy demand in Iran, and for this reason the
present chapter is confined for the most part to the study of these
four products.

RELATIONSHIP BETWEEN GNP AND
DEMAND FOR ENERGY

The relationship between the level of economic development
and energy consumption is well established, and numerous publications on the subject are available. It is generally agreed that there
is a close correlation between the rate of economic growth and the
demand for energy. Professor Mikdashi has correlated per capita
GNP and per capita energy consumption for the United States, Northern Europe, Southern Europe and Japan, and Latin America for
1960-66. He obtained a linear relationship of $Y = 0.14$, $X - 2.5$;
$r^2 = 0.996$, where X is per capita GNP and Y per capita energy consumption.[1] The Stanford Research Institute has also carried out a
project examining the relationship between energy demand and GNP,
for Iran, Portugal, Greece, Turkey and Brazil, Japan, and the
United States. In all these cases it reported that a definite and
direct relationship exists between the rate of growth of GNP and the
rate of growth of demand for energy. The relevant data for Iran are
shown in Table 10.3.

The figures in Table 10.3 show three distinct periods:

(a) 1960-65: The Iranian recession of 1960-63 affected energy
consumption more than GDP as a whole. In the 1964-65
period the economy showed signs of recovery and the energy
and GDP growth rates began to converge. Generally speaking, there were too many distortions in this period to allow
proper analysis, but it is fair to say that toward the end of
this period industrialization got under way at a fast rate.

(b) 1965-69: The growth of energy demand overtook that of
the GNP, indicating rapid economic progress and the
greater utilization of energy resources.

(c) 1969-82: The growth of energy demand and GNP run parallel to each other. This is a pattern similar to that of the industrial countries of the world, and may indicate that Iran has started a phase of self-sustained growth.

TABLE 10.3

Annual Growth Rates of GNP and Energy Demand
(percents)

	1960-65	1965-69	1969-82*
Gross national product	6.70	8.60	11.80
Energy demand	6.30	11.50	11.20
Ratio of energy growth to GNP growth	0.94	1.34	0.95

*Projected GNP growth rate is based on the Plan Organization forecasts in 1969. Projected energy demand growth rate is based on SRI forecasts in 1969.

Source: The Plan Organization.

In the case of Iran, one could argue that there is a strong and direct correlation between the rate of economic growth and the consumption of the four main products which constitute the bulk of the Iranian energy demand.[2]

HISTORICAL MOVEMENTS IN THE DEMAND
FOR OIL PRODUCTS

The analysis of the historical movements in demand for oil products has a twofold advantage. First, we are able to examine the factors behind the variations in demand and this will help to forecast the future movements of the demand; second, we can obtain an indication of the level of economic development in the country by looking at the various trends. Table 10.4 shows the demand for oil products in Iran since 1950. The growth rate of demand was 10.3 percent in 1961, and 11.8 percent in 1974, indicating the persistent demand for oil products in a decade of rapid industrialization.

Factors Affecting the Demand

Lack of reliable data has made it impossible to make a detailed analysis of all the movements in demand. Therefore we confine our analysis to two particular areas: first, the general factors affecting the oil product demand in Iran; and second, the specific events that have influenced this demand. The general factors are as follows.

(a) General economic growth and industrialization. The government policy and the way it affects the direction of the economic growth had an important influence on the demand for oil products. The Third Development Plan (1963-67) anticipated industrialization on a massive scale. The rate of growth of GNP was 8.5 percent and the rate of growth of industry was 11 percent, while the oil product demand grew by 13 percent. In the Fourth Development Plan (1968-72) the rate of growth of GNP was 11.4 percent and that of the oil product demand 10.3 percent.[3]

(b) Expansion of the distribution and transport network (Chapter 9).

(c) Stability of the political system, attracting foreign and domestic capital to set up industries which were naturally petroleum-based.

(d) Population increases. The population of Iran was 17.5 million in 1950, and over 33.5 million in 1974.[4] While the Iranian population has less than doubled in 24 years, the consumption of oil has risen seventeenfold. Indeed, oil consumption per head was 120 liters in 1957, and 219 liters in 1964. In 1974 this figure rose to 510 liters.[5]

(e) Urbanization. Since the urban population consumes the largest quantity of petroleum products, the growth in urbanization must affect the demand for oil products. According to the 1956 National Census, the total number of centers with a population of more than 5,000 was 186. In the 1966 Census this figure reached 235. The Iranian urban population was 31 percent of the total in 1951, but rose to 43 percent in 1970. In the 1956-70 period the urban population growth was 5.3 percent per annum, while the rural population grew by only 1.7 percent per annum. The average annual growth of population in this period was 2.9 percent per annum.[6]

(f) Per capita income. Increases in per capita income affect the demand for various oil products differently. In some cases the demand is affected directly through the expansion of the market. Thus, for example, in the case of gasoline, the increase in per capita income will induce new consumers to enter the market but will not significantly affect the consumption level of those already in the market, unless they purchase additional motor cars. In other

TABLE 10.4

Demand for Various Oil Products in Iran for Selected Years
(thousands of cubic meters)

	1950	1955	1957	1958	1959	1960	1961	1962	1963	1964
LPG	--	--	1	1	3	5	8	12	15	24
Aviation fuel	--	--	29	38	36	33	24	22	23	25
Jet fuel J.P. 4	--	--	3	14	29	53	24	30	31	37
Jet fuel A.T.K.	--	--	--	--	--	--	63	80	84	93
Total aviation fuel	--	--	32	52	65	86	111	132	138	155
Gasoline	246	383	464	524	575	627	644	664	702	740
Kerosene	271	524	713	752	886	976	1,099	1,171	1,258	1,436
Gas oil	84	235	385	520	707	856	984	1,083	1,158	1,349
Fuel oil	422	796	828	874	1,027	1,147	1,222	1,259	1,268	1,567
Total: four main products	1,023	1,948	2,390	2,670	3,195	3,606	3,949	4,177	4,386	5,032
Iranol lubricating oil	--	--	11	13	15	16	18	21	23	30
Tar	--	--	41	60	78	134	157	91	73	104
Other products	--	--	8	10	11	15	15	19	29	42
Total consumption	1,023	1,948	2,483	2,806	3,367	3,862	4,258	4,452	4,664	5,387

	1965	1966	1967	1968	1969	1970	1971	1972	1973	1974
LPG	31	46	70	104	148	201	254	319	305	--
Aviation fuel	27	34	35	32	29	22	18	9	10	--
Jet fuel J.P. 4	40	47	49	54	96	126	151	181	239	--
Jet fuel A.T.K.	101	110	156	149	158	215	266	303	386	--
Total aviation fuel	168	191	240	235	283	363	435	493	635	--
Gasoline	714	800	854	1,943	1,068	1,227	1,413	1,598	1,928	2,304
Kerosene	1,470	1,530	1,808	1,990	2,336	2,290	2,514	3,099	3,244	3,730
Gas oil	1,593	1,881	2,185	2,467	2,654	3,016	3,364	3,648	4,522	5,065
Fuel oil	1,712	1,979	2,238	2,577	2,820	3,146	3,319	3,378	4,024	4,429
Total: four main products	5,489	6,190	7,085	7,977	8,878	9,679	10,610	11,723	13,718	15,528
Iranol lubricating oil	35	39	43	50	55	61	71	76	96	--
Tar	126	159	201	206	262	257	357	345	433	--
Other products	46	41	48	71	84	91	109	114	148	--
Total consumption	5,889	6,672	7,687	8,643	9,710	10,652	11,836	13,070	15,425	17,237

Source: Statistical Office of the Distribution Department.

cases--for example, kerosene--increases in per capita income will
influence the demand in two ways. On the one hand it enables consumers to purchase additional space heaters or new facilities such
as water heaters. On the other hand, the increases in per capita
income may cause a switch by other consumers from coal and wood
to kerosene. Little work has been done in Iran on estimating the income elasticity of demand. The only recorded result was published
by the Central Bank of Iran, on the basis of average data for ten
years. These data are shown in Table 10.5. The data in this table
seem to confirm the argument that the increase in per capita income
has a smaller effect on the consumption of motor car fuels. The income elasticity of demand for fuel oil is relatively high, which is
not surprising in a rapidly growing economy. It may be interesting
to note that a United Nations study in 1968 showed that the average
income elasticity coefficient for energy consumption was 1 for developed private enterprise economies; 1.2 for centrally planned
economies, and 1.6 for developing countries.[7]

TABLE 10.5

Income Elasticity of Demand for Various
Oil Products in Iran, 1958-68

Product	Income Elasticity of Demand
Gasoline	0.72
Kerosene	1.40
Gas oil	0.80
Fuel	1.70
Other products	4.37
Total	1.82

Source: Bank Markaz, Iran, "Annual Report and Balance
Sheet, 21st. March 1968." (In Persian.)

(g) Price changes. Because of the relatively steady prices
maintained by the National Iranian Oil Company, no calculation of
the price elasticity of demand has been possible. However, it is
generally agreed that the petroleum product demand is price-
inelastic in Iran. The price elasticity of demand is, to a great extent, dependent on the availability and price of substitutes. In the
early 1960s Professor Adelman noted that the price elasticity of

demand for crude oil was "a composite of the moderate gasoline
elasticity, the very great fuel oil elasticity and the intermediate one
for middle-distillates."[8] Prices of fuel oil and coal were then competitive, thus making for a high cross-elasticity of demand. As the
price of fuel oil dropped and that of coal increased, the price elasticity of demand for fuel oil declined. The price elasticity of fuel
oil could increase once nuclear power or natural gas become competitive. In Iran, however, coal was never a serious competitor to
fuel oil; wood and charcoal were inferior to kerosene, and not available in large quantities. Gasoline and automotive gas oil were naturally competitive as motor car fuels, but it is not expected that a
change in their relative prices will cause a switch from one product
to the other. The government has deliberately imposed high taxes
on gasoline, while making it illegal to use diesel-driven motor
vehicles inside the cities. In this way the government has subdivided
the transport market into two distinct areas: (i) private motor cars,
taxis, and public transport, using gasoline only, may be used in
the cities; (ii) diesel-driven motor vehicles, such as lorries, inter-
city bus services, trains, road tankers, and tractors, can be used
outside the urban areas. This separation of markets has eliminated
the cross-elasticity of demand between automotive gas oil and gasoline.

Having discussed the general factors that affect the demand for
oil products, let us now consider some specific events in Iran which
have had a profound impact on the demand.

1960-63 Economic Recession

This period was marked by a severe economic crisis, a slowdown in industrial activity combined with the drought of 1962, which
adversely affected agricultural production. The effect of this economic recession can be seen from Table 10.4. The rate of growth
of all products dropped from 10.3 percent in 1961 to 4.6 percent and
4.8 percent in 1962 and 1963. Although there was no decline in the
rate of growth of the demand for gasoline, this growth was small.
The rate of growth of kerosene and gas oil fell sharply. The largest
decline was that of fuel oil, the most important industrial fuel; its
growth in 1962 was half that of 1961. The rate of growth of GNP fell to
2 percent in 1962--the lowest annual increase since the nationalization.

1964 Price Increase

In 1964 the Cabinet decided to raise the prices of gasoline and
kerosene, upon the recommendation of the Board of Directors of NIOC.

Gasoline prices were increased by 100 percent from 5 rials per liter
to 10 rials per liter, and kerosene prices by 48 percent from 2.3
rials to 3.5 rials per liter. Prices of fuel oil and gas oil remained
unchanged. The government's decision to increase prices was based
on two factors--first, domestic prices were very low compared to
external prices,[9] and second, no public reaction was anticipated,
given the extent of dependence on these two products; that is, the
price elasticity of the demand was thought to be negligible. The
government had clearly misjudged the situation. The public reaction was violent and spontaneous. A boycott of gasoline consumption
started immediately. Taxi drivers and bus drivers went on strike
and the private motorists stopped using their cars. There were
large demonstrations by workers, students, and the middle class.
The Shah personally intervened and ordered the reduction of gasoline and kerosene prices to 6 rials and 2.5 rials per liter respectively. Following the increase in prices, there was a decline in the
absolute level of demand for gasoline, while the rate of growth of
the demand for kerosene fell from 14.2 percent in 1964 to 2.3 percent in 1965. After the reduction in prices the demand returned to
normal.

The peculiarity of the situation, where we have an already low
price for petroleum products, and a rise in prices causes such a
public reaction, can only be explained in terms of psychological factors. It is not the price elasticity of demand that we should look for,
but rather the political elasticity of demand. This stems from a
strange kind of emotion and anger which is associated with oil in
Iran. It seems reasonable to assume that the emotions over oil were
an aftermath, if not an extension, of the Iranian nationalization and
its subsequent defeat. There is, even today, a feeling among many
Iranians that they have been cheated out of their rightful share of
this God-given resource, and they simply are not prepared to pay
more for it. Indeed, the oil product prices have been frozen since
1965.

DEVELOPMENT OF OIL MARKETS IN IRAN

The examination of various end-use markets for petroleum
products in a country can lead not only to a better understanding of
the sectoral demand for those products, but also to a realization of
the state of development of the economy. It will indicate that the
consumption of oil products has been both the cause and effect of
the increased level of economic activity.

It is possible to distinguish ten major internal markets in
Iran, as shown in Table 10.6. It can be seen that three sectors

TABLE 10.6

End-Use Analysis of Petroleum Demand in Iran for Selected Years

(thousands of cubic meters)

	1960 Quantity	1960 Percent	1965 Quantity	1965 Percent	1969 Quantity	1969 Percent	1974[a] Quantity	1974[a] Percent	Growth Rate per Annum, 1969–74, Percent
Residential and commercial	1,501	40	2,137	37	3,260	34	5,258	31	10
Electricity	234	6	352	6	728	8	1,350	8	13
Industrial	790	21	1,269	22	2,143	23	4,584	27	16
Government	54	1	116	2	220	2	340	2	9
Agriculture	51	1	283	5	606	6	1,350	8	17
Road transport	895	24	1,273	22	1,963	21	3,396	20	12
Railways	79	2	86	1	93	1	170	1	13
Aviation	86	2	168	3	282	3	340	2	4
Coastal shipping	--	--	3	--	8	--	15	--	14
Oil company use	100	3	142	2	166	2	177	1	1
Total end-use	3,790	100	5,829	100	9,469	100	16,980	100	13
Other[b]	72	--	60	--	241	--	257	--	--
Total internal demand	3,862	--	5,889	--	9,710	--	17,237	--	13

[a]1974 figures are estimates.

[b]Petroleum consumption outside the above markets.

Note: For a detailed breakdown of the oil markets and the regional importance, see Stanford Research Institute's Report to the Government of Iran, Long Term Energy Policy in Iran, Vols. II and V.

Source: NIOC for data, SRI for classification.

predominate: residential and commercial, industrial, and road transport; together they accounted for 78 percent of the demand in 1974. Historically, the residential and commercial market has declined in importance, and this trend is expected to continue in the future, mainly because of the introduction of natural gas. Industrial market has improved its position, while road transport has slightly declined in importance. The latter is, however, expected to become the largest end use by 1982. Agricultural has substantially improved its relative position from 1 percent in 1960 to 8 percent in 1974; other end users have more or less maintained their relative positions.

Insofar as the growth rates are concerned, we can see that in most of this period agriculture has shown the highest growth rates as a result of improvements in methods of production and possibly the land reform. The industrial market using fuel oil and gas oil has maintained a growth rate of 10-16 percent throughout this period, closely followed by road transport. On the regional basis, the industrial users of petroleum in the Tehran region consumed twice as many oil products than the average for the country, while Kerman's consumption of industrial fuels is less than half the average for the country, indicating the level of regional imbalance in the industrialization of the country.

THE ROLE OF OIL IN HOUSEHOLD EXPENDITURE

One way of assessing the demand for various oil products in Iran is to consider the expenditure of households on these products. The only set of data available is the Urban Household Budget Survey of Iran, prepared by the Central Bank of Iran.[10] The data used in this section are for 1967 and 1969, showing the expenditure on the main sources of energy and its comparison with other major categories of expenditure. (See Tables 10.7 and 10.8.) With regard to the 1967 figures, we can make the following observations:

1. The oil product consumption is directly proportional to the level of income; unlike many other sources of energy, the absolute expenditure on oil products does not stabilize or decline after a certain level of income is reached. Indeed, there is a 25 percent increase in the expenditure between the 300,000-400,000 rials income group and the 400,000- 500,000 rials income group (the two highest income groups).
2. Generally, the expenditure on coal is a minor item at all levels. However, while the expenditure has been declining among higher income groups (through the substitution of oil products) there has been a steady rise in the lower income groups.

3. Gas consumption in Iran is confined to LPG used for cooking by the urban middle class. The only large city in Iran which uses natural gas through pipelines is Shiraz. The two lowest income groups consume no gas at all, while this expenditure is very small in the higher income groups.

4. Wood consumption is very much a poor man's means of cooking and heating. The expenditure on wood remains constant for the first few income groups, but falls steadily as income increases.

5. Electricity consumption very much follows the pattern set by the oil products. Expenditure is directly related to the increases in the level of income.

6. In terms of average expenditure, oil products are in the first position, followed by electricity and then coal. Wood and gas consumption occupy fourth and fifth positions respectively.

The 1969 data are better indications of household expenditure, since the sample size is larger. The expenditure on oil products and electricity is still the highest. However, the ratio of the expenditure on oil products to that of electricity altered from 2.5 in 1967 to 1.6 in 1969. In the highest income group, the expenditure on electricity overtook that of the oil products for the first time. Gas consumption expanded, and it was used even in the lowest income groups. It is expected that toward the end of the 1970s, oil product expenditure will still be in first place, followed closely by gas and electricity expenditures. Coal will be in fourth place and wood will be in last place.

Comparison of the Oil Product Expenditure with Other Main Categories

The Central Bank of Iran has divided the total household expenditure into ten categories. Table 10.8 provides an account of these categories for the various income groups in 1969. Each category is composed of between 10 to 70 items. For instance, the oil product expenditure is only one of the 19 items in category 3--the running cost of the household.

The comparison is useful insofar as it points out that in the lower income groups the expenditure on oil products is often larger than the expenditure on some main categories. Thus the following observations can be made:

1. Oil product expenditure, on average, constitutes 25 percent of the total expenditure in category 3.

TABLE 10.7

Importance of Energy Expenditure in Various Households, 1967 and 1969

Income Groups Consumers and Energy Products	0-30,000	30,000-39,999	40,000-49,999	50,000-74,999	75,000-99,999	100,000-149,999	150,000-199,999
1967							
Number of households	320	173	150	310	222	202	94
Average number of people in a household	3.53	4.50	4.96	5.08	5.67	5.69	5.55
Coal	353.9	388.6	424.6	445.5	460.4	485.3	388.5
Oil products	979.0	1,413.1	1,594.6	1,761.1	1,932.9	2,336.5	2,813.8
Gas consumption	--	--	3.1	1.9	82.5	116.0	415.6
Wood consumption	168.9	135.0	162.3	207.0	163.3	154.2	349.9
Electricity	166.3	294.5	413.9	563.3	887.6	1,303.2	1,824.6
1969							
Number of households	380	686[a]		763	630	863	509
Average number of people in a household	2.67	4.06[a]		4.93	5.06	5.56	5.84
Coal	165.8	302.8[a]		477.4	591.3	754.3	769.7
Oil products	739.8	1,461.3[a]		1,796.8	2,062.5	2,450.9	2,886.9
Gas consumption	3.8	10.6[a]		42.5	109.4	218.9	386.7
Wood consumption	98.4	130.7[a]		139.7	129.6	148.9	407.2
Electricity	75.8	403.2[a]		503.5	798.9	1,429.4	1,714.7

Income Groups Consumers and Energy Products	200,000–249,999	250,000–299,999	300,000–399,999	400,000–499,999	500,000 and over	Average Weighted Expenditure
1967						
Number of households	41	30	36	12	23	--
Average number of people in a household	6.32	6.20	6.08	6.33	6.35	4.98
Coal	421.3	366.3	293.6	387.5	331.1	415.9
Oil products	3,312.4	3,641.1	4,459.0	5,750.8	7,312.0	1,954.3
Gas consumption	544.0	692.0	1,030.0	1,866.7	2,302.6	146.7
Wood consumption	212.0	52.0	357.5	113.3	78.3	175.0
Electricity	2,598.8	2,514.4	3,196.8	4,370.8	4,840.4	887.1
1969						
Number of households	468[b]		195	96	193	--
Average number of people in a household	5.97[b]		5.96	6.19	6.19	5.07
Coal	631.8[b]		1,003.8	921.5	248.7	553.3
Oil products	3,383.4[b]		4,078.6	4,538.3	6,561.5	2,417.0
Gas consumption	600.9[b]		1,047.2	1,341.3	1,959.1	310.8
Wood consumption	148.9[b]		317.8	78.9	59.3	165.9
Electricity	3,125.5[b]		3,321.4	4,024.9	7,149.4	1,496.2

[a]From 30,000 to 49,999.

[b]From 200,000 to 299,999.

Source: Bank Markaz Iran, unpublished data provided by the Economic Statistics Department, Household Budget Survey section.

TABLE 10.8

Summary of the Main Categories of Expenditure in Comparison with
Oil Products Expenditures, 1969

(rials p.a.)

Income Groups Items	0– 30,000	30,000– 50,000	50,000– 75,000	75,000– 100,000	100,000– 150,000	150,000– 200,000	200,000– 300,000	300,000– 400,000	400,000– 500,000	500,000 and over	Average Expenditure
Expenditure on oil products	739.8	1,461.3	1,796.8	2,062.5	2,450.9	2,886.9	3,384.4	4,078.6	4,538.3	6,561.5	2,417.0
1. Food	9,845.2	20,941.3	28,933.0	37,970.4	51,396.6	65,354.0	84,084.0	121,797.3	138,508.9	174,110.8	52,352.7
2. Accommodation	873.4	2,029.2	3,583.8	5,140.6	7,603.1	15,156.4	23,410.4	33,025.5	50,661.5	175,364.8	16,436.9
3. Running cost of the household	2,149.8	3,992.3	5,236.0	6,615.0	9,173.8	11,052.6	15,462.3	19,604.2	26,543.7	34,183.5	9,229.9
4. Traveling expenditure	273.6	833.9	1,548.5	2,401.6	3,746.9	5,882.0	11,242.4	18,919.2	42,996.0	93,442.6	8,582.1
5. Furniture, etc.	364.5	782.4	2,267.0	4,408.4	6,021.9	11,716.5	15,784.8	25,967.7	24,388.4	46,170.3	8,341.7
6. Personal expenses	743.1	1,682.6	2,525.1	3,406.1	4,145.2	5,277.8	6,009.0	7,469.5	9,343.1	12,031.9	3,998.4
7. Health	495.6	1,577.0	2,703.8	3,625.9	5,858.6	7,806.1	11,164.7	11,639.9	15,292.2	42,069.4	6,621.7
8. Education	133.3	380.6	743.5	1,163.6	1,785.9	3,084.4	4,730.7	7,530.4	9,930.0	16,099.5	2,604.1
9. Miscellaneous	944.1	2,274.9	3,423.0	5,071.1	7,427.9	10,787.6	16,230.9	25,432.3	30,686.9	62,247.2	9,854.0
10. Clothing	1,310.0	3,969.3	6,628.2	9,676.6	13,809.1	20,617.7	26,264.4	36,562.6	41,463.5	65,882.3	15,162.5
Expenditure on oil products as a percentage of total	4.3	4.3	3	2.7	2.1	1.8	1.5	1.2	1.1	0.9	1.8

2. In the first income group, the expenditures on oil products and accommodation are nearly equal.
3. Expenditure on household furniture is lower than those on oil products for incomes under 50,000 rials per annum.
4. Expenditure on health is lower than that on oil products for the first income group, while in the second income group the two are very close.
5. Expenditure on education is lower than oil product expenditure in the first five income groups.

For the large majority of Iranians who are in the lower income groups, oil products expenditure constitutes an important expense. Given the dependence on oil products, a rise in petroleum product prices would cause a great deal of hardship for the poorer population. Since the larger part of the expenditure on oil products is related to kerosene, this analysis could provide the government with a strong case for the stabilization of the oil product prices in general and that of kerosene in particular.

SUMMARY AND CONCLUSION

There is clearly a close relationship between the level of economic development and the demand for energy in all the countries of the world. How close this relationship may be depends on the stage of the industrial development of that country.

Oil has been playing an increasingly important role in supplying the energy requirements of the industrial countries of the free world. Indeed, these countries draw at least 45 percent of their energy demand from oil. In 1969 oil products accounted for 74 percent of the energy consumption in Iran, a ratio similar to that in Japan. The four main products--gasoline, kerosene, gas oil, and fuel oil--together supplied over 90 percent of the oil product demand, acting as an important indicator of the overall energy demand in Iran. Among the most significant factors contributing to rise in demand are government strategy on industrialization and economic growth, expansion of domestic distribution network, population increase, urbanization, and increase in per capita income. Price elasticity of demand was argued to be negligible, but the price increases of 1964 which resulted in sharp decline in demand were attributed to a "political elasticity of demand" rather than price considerations.

Ten major internal markets for petroleum products were discussed in this chapter, among which three markets predominated-- residential and commercial, industrial, and road transport. The relative position of these markets and their rate of growth indicates the way the end-use markets have grown themselves. Indeed, a close study of the development of these individual markets will clearly reflect the changes in the government strategy as well as the impact of various policies on the demand for products by these end users.

Finally, the importance of oil product expenditure in the household budget was examined. The conclusions were that the expenditure on oil products was larger than those on electricity, coal, gas, and wood. In the lower income groups the expenditure on oil products exceeded that on main categories such as health and education. The importance of the oil product expenditure in the household budget has far-reaching policy implications for the government insofar as it strengthens the case for maintaining low prices for oil products rather than allowing economic prices to be charged.

NOTES

1. Z. Mikdashi, The Community of Oil Exporting Countries (London: Allen & Unwin, 1972), pp. 120-21. See also S. Schurr and P. T. Homan, Middle Eastern Oil and the Western World--Prospects and Problems (New York: Elsevier Publishers, 1971), p. 172, and Our Industry, Petroleum (London: British Petroleum Company Ltd., 1970), chap. 1.

2. A regression was run for Iran's GDP and the demand for the four main products for 1957-71. The regression line was $D = -164.58 + 6.55 \, GDP$, with $r^2 = 0.99$. The very high correlation coefficient is not surprising in view of the close interrelationship between the variables.

3. GNP figures are taken from The Annual Report and Balance Sheet, Central Bank of Iran. Oil product demand growth rates are calculated from Table 10.4. Note: These figures are not strictly comparable to those in Table 10.3, as the latter considers the demand for all energy resources.

4. J. Bharier, Economic Development in Iran, 1900-1970 (London: Oxford University Press, 1971), p. 27. The Iranian National Census was carried out in 1966. All the figures are based on a 3 percent rate of growth of population.

5. NIOC Statistical and Information Office.

6. Bharier, op. cit., p. 28.

7. UN Economic and Social Council, Natural Resources Development and Policy (New York: United Nations, 1971), pp. 5-6.

8. M. A. Adelman, "The World Oil Outlook," in Natural Resources and International Development, edited by M. Clawson (Baltimore: Johns Hopkins Press, 1964), p. 84.

9. The price of industrial gas oil in Iran is equal to that of automotive gas oil (diesel oil). See Chapter 12.

10. The survey started in the mid-1950s, but was discontinued in 1960. The second series started in 1964 and was carried out for households of 35 large and small towns and cities in Iran. No comparable work has been undertaken for rural areas. Due to the continuous change in the composition of these surveys, recent comparative data could not be used.

11

THE IMBALANCE BETWEEN
SUPPLY AND DEMAND
FOR OIL PRODUCTS

In Chapter 9 we referred to the imbalance between the supply and demand for oil products as a major problem facing NIOC. The present chapter will examine this problem in depth and attempt to offer remedies for its solution.

The imbalance has come about because of the peculiar way in which the demand for oil products has been growing in Iran during the past decade. Indeed, the demand for the middle distillates has constituted over 50 percent of the total demand for oil products since 1965. We know that the refinery pattern dictates the product mix, which is relatively inflexible, and any changes in its pattern will be small in the short run. Thus increases in the level of demand will consistently widen the gap between the supply and demand of some oil products.

In the oil-consuming countries, where the refining and distribution of oil products is dominated by the major oil companies, the problem of imbalance has been easily overcome. The oil companies simply switch the excessive quantities of some oil products to the countries that suffer a shortage of these products, and vice versa. However, the problem of imbalance has been particularly acute in Iran because of (a) the great dependence of the Iranian consumers on kerosene for heating, cooking, and lighting (in villages), (b) government policy, which has encouraged the use of diesel oil rather than gasoline, (c) the difficulty of exporting oil products,[1] and (d) the political inexpediency of importing certain refined products,[2] coupled with its high cost, which has meant that NIOC was unable to balance its supply and demand in the same way that the international oil companies had been doing for some time.

At present there are four refineries and a topping plant operating in Iran--Tehran, Shiraz, and Kermanshah refineries, MIS

topping plant, and the Abadan export refinery. The management of
the latter has been passed by the Consortium to NIOC under the 1973
Agreement, yet for all practical purposes the Consortium's share of
the liftings has not changed. Table 11.1 shows the refinery pattern
in Iran (excluding the Abadan export refinery).

TABLE 11.1

Refinery and Demand Pattern for Oil Products in Iran, 1974
(percents)

Product	Refinery Yield (1)	Demand Pattern (2)	1 - 2
Gasoline	18	13	+5
Kerosene	8.5 } 33	22 } 51	} -18
Gas oil	24.5	29	
Fuel oil	40	26	+14
Other products	9	10	-1
Total	100	100	0

Source: Chapters 8 and 10.

NIOC uses the output of the Tehran and Kermanshah refineries
for domestic purposes. It is also empowered to lift up to a maximum
of 100,000 b/d (5.8 million cubic meters a year) of oil products from
the Abadan refinery and some fuel oil from the MIS topping plant, if
and when required.

While NIOC is obliged to lift a proportion of all the oil products
refined at Abadan, it is able to change the composition of the products
lifted after negotiations with the Consortium. Based on its internal
forecasts, NIOC produces every year a schedule of its proposed liftings from Abadan to the Consortium. Although the product mix of
these liftings may cause the Consortium a loss of export business,
it almost always has given way to NIOC's demands. The Consortium
has made it known that it does not wish to involve itself in a quarrel
with NIOC over a relatively small product export for fear of jeopardizing its larger dealings in crude. Indeed, NIOC has never lifted its full
entitlement of 5.8 million cubic meters, simply because it did not
need to do so. Thus, NIOC has been lifting oil products from Abadan
in such a way as to balance the domestic supply with demand. Table
11.2 illustrates this point. The Abadan refinery accounted for 45

TABLE 11.2

Supply Pattern of the Iranian Refineries, 1974
(thousands of barrels)

	Tehran	Kermanshah	MIS	Shiraz	Abadan*	Total	Abadan as a Percent of Total
Gasoline	5,726	1,269	--	2,063	5,900	14,958	39
Kerosene	5,733	1,029	--	1,671	12,600	21,033	60
Gas/diesel oil	9,923	763	--	2,250	19,200	32,136	60
Fuel oil	11,775	2,873	9,462	5,635	981	30,726	3
Other	890	278	--	669	12,100	13,937	39
Total	34,047	6,212	9,462	12,288	50,781	112,790	45

*Reflects the portion delivered to NIOC for domestic consumption.

Source: Statistical and Information Office of the Affiliated Companies of NIOC.

percent of the domestic refinery output in Iran in 1974 but the proportions of kerosene and gas oil liftings from Abadan provided for 60 percent of the domestic demand for these products.

The Consortium Annual Reports show that in 1974, 89 percent of the fuel oil and 59 percent of gasoline produced at Abadan and MIS were lifted for exports. In the same year NIOC lifted 93 percent of kerosene and 69 percent of gas oil produced by Abadan for internal use.

The use of Abadan as a "swing" refinery raises an important question: How long can Abadan continue to fill the gap between the supply and demand for middle distillates in Iran? This will be the subject of discussion on short-term and long-term solutions to the problem of product imbalance in Iran.

SHORT-TERM AND LONG-TERM MEASURES

NIOC has the following options to deal with the problem of imbalance between the supply and demand for the middle distillates:

(a) Take over the Abadan refinery and use all its middle distillates for the domestic demand, forgoing all export profits on these products.
(b) Import the required middle distillates.
(c) Build new refineries with costly isomerization units which produce larger proportions of middle distillates.
(d) Speed up the plans for natural gas distribution.
(e) Restructure the pricing system.

The first two are short-term measures, the others long-term alternatives.

SHORT-TERM SOLUTIONS

NIOC has in the past used Abadan as a "balancing" refinery and no doubt will continue to do so in the future. As the demand for oil products increases, it will become more and more difficult for NIOC to balance its supply and demand through Abadan. Although NIOC is theoretically in full control of the Abadan refinery, its relationship with the former consortium is basically unchanged under the 1973 Sales and Purchase Agreement. Thus NIOC's liftings will be limited to its past entitlement, but NIOC will no doubt seek to modify the agreement if it finds it necessary to do so. Yet even if NIOC proceeds to lift all the middle distillates produced at Abadan, it will still not be able to satisfy the domestic middle-distillate demand (see Table 11.3). The use of Abadan as a "balancing" refinery will be short-lived, and other measures will have to be introduced to alleviate the situation. Import of middle distillates could be the simplest short-term solution. However, this possibility has been ruled out for two reasons--the political implications of such a move, and the high price of middle distillates in the international markets.

LONG-TERM SOLUTIONS

New Refineries

NIOC's past record with regard to the construction of refineries has been poor. The Shiraz refinery--the second NIOC-built domestic refinery in more than 20 years--with a capacity of 40,000 b/d--was scheduled for completion in 1971, but did not become operational until November 1973. By early 1973, NIOC had found itself in a difficult situation. Previous forecasts for domestic demand had been widely off mark and the new forecasts indicated an unprecedented growth in demand particularly for middle distillates throughout the 1970s. Thus, in a short period a wave of refinery construction and feasibility study started. Contracts were awarded to foreign firms for building the necessary equipment, and NIOC's refinery division was busily planning the designs. In March 1975 the second Tehran refinery was inaugurated, while three other refineries in Tabriz, Esfahan, and Neka are expected to come onstream by 1978. The past failure of NIOC to construct (or carry out feasibility studies for) refineries ahead of time stems from the fact that the company had never faced a shortage of domestic refining capacity in the past and had consequently underestimated the seriousness of the potential problem. Although this lack of foresight has now been partially compensated, it must, however, be emphasized that while the construction of new refineries might help to balance the middle-distillate shortage, it will at the same time provide excessive quantities of light and heavy distillates, which cannot easily be marketed abroad.

The Role of Gas

Since the discovery of oil in Iran, the larger part of the associated gas has been flared. In 1965 an Irano-Soviet agreement was signed for the export of gas to Russia. The IGAT Project (Iranian Gas Trunkline) was completed in 1970. IGAT was designed to serve three purposes:

TABLE 11.3

Estimates of Supply and Demand for Middle Distillates, 1974–82
(millions of barrels)

	Kerosene					Gas/Diesel Oil				
	1974	1976	1978	1980	1982	1974	1976	1978	1980	1982
Supply[a]										
Abadan[b]	12.6	12.6	12.6	12.6	13.5	19.2	19.2	19.2	28.0	28.0
Kermanshah	1.0	1.0	1.0	1.0	1.0	0.8	0.8	0.8	0.8	0.8
Shiraz[c]	1.6	3.9	3.9	3.9	3.9	2.2	2.9	2.9	2.9	2.9
Tehran I	5.7	5.7	5.7	5.7	5.7	10	10	10	10	10
Tehran II[d]	--	4.8	4.8	4.8	4.8	--	10	10	10	10
Tabriz[e]	--	6.6	6.6	6.6	6.6	--	8.7	8.7	8.7	8.7
Esfahan[e]	--	11.2	11.3	11.3	11.3	--	--	17.5	17.5	17.5
Total supply[f]	20.9	34.6	45.9	45.9	46.8	32.2	51.6	69.1	77.9	77.9
Total demand[g]	20.8	32.0	37.2	44.2	51.6	31.9	47.0	61.0	79.2	103.0
Excess/shortage	0.1	3.6	8.7	1.7	-4.8	0.3	4.6	8.1	-1.3	-25.1

[a] All refineries are assumed to operate at maximum capacities.

[b] NIOC is assumed to lift the entire middle-distillage output of Abadan whenever shortages occur, as in 1980 for gas oil and 1982 for kerosene.

[c] Shiraz refinery is assumed to operate at full designed capacity by 1976.

[d] Tehran second refinery started operation in March 1975.

[e] Tabriz and Esfahan refineries are expected to come onstream in 1976 and 1978 respectively.

[f] A 130,000 b/d refinery is also to be constructed by 1977/78 at Neka--no details are as yet available.

[g] Based on the forecast by the Distribution Department in 1973 (unpublished).

Source: Compiled by the author.

(a) To earn extra foreign exchange by exporting gas to the Soviet Union.
(b) To bring about closer economic and political cooperation between Iran and her old enemy Russia.[3]
(c) To supply the Iranian market with gas along its 1,100-km route to the Soviet border.

Ironically, the supply of gas for domestic consumption, which was the least important factor in the construction of IGAT, has become an integral part of NIOC's planning for the future oil product demand in Iran. Indeed, until 1968 no serious discussion of the role of gas took place in official circles, which indicated the absence of an over-all energy policy.

Gas is a substitute for fuel oil, industrial and heating gas oil, LPG, and kerosene, but gas consumption will not immediately ease the middle-distillage problem. The most immediate and profound impact of gas will be on fuel oil. This is because the industrial concerns are concentrated along the IGAT route and no extensive branch pipeline system is required. Later, toward the end of the 1970s and early 1980s when the NIOC plans for intracity gas pipelines are completed, the supply of gas will help to ease the middle-distillate shortage. Table 11.4 shows the expected impact of natural gas on the domestic demand for oil products.

TABLE 11.4

Expected Impact of Natural Gas on
Demand for Oil Products
(thousands of cubic meters)

Oil Products Substituted by Natural Gas	1972	1978
LPG	--	95
Kerosene	3	393
Gas oil	--	315
Fuel oil	1,813	4,185

Source: J. Zamanian, The Role of Oil Products in the Security of IRAN, NIOC Public Relations Office, 1971, p. ?8 (in Persian).

The above table shows that in the short-term natural gas will replace fuel oil and will thus not help to reduce the demand for middle distillates. In the long term, however, natural gas will replace kerosene and gas oil.

A Change in the Structure of Prices

Rather than accommodating the demand by building refineries with costly isomerization units (to increase middle-distillate yield), prices can be used to bring the demand pattern more in line with the supply pattern. (This is dealt with in Chapter 12, Summary and Conclusion.)

SUMMARY AND CONCLUSION

The imbalance between the refinery pattern and the demand for oil products in Iran has become particularly important since the mid-1960s. NIOC has traditionally used the Consortium-operated Abadan refinery as an adjustment mechanism for correcting the supply bottlenecks. But as the rate of consumption increased, it became clear that this mechanism had outlived its usefulness. At the same time, unbalanced liftings from Abadan have cost NIOC a considerable sum of foreign exchange in terms of forgone export revenues, due to the high prices of middle distillates.

Five alternatives were offered in this chapter to ease the middle-distillate problem. These were (a) takeover of the Abadan refinery's total middle-distillate output, (b) imports of middle distillates, (c) construction of new refineries, (d) the use of natural gas, and (e) a change in the product prices. The first alternative was found to serve as a temporary measure only, while the second (and perhaps the simplest) alternative was ruled out for political, as well as economic reasons. NIOC's performance with regard to the construction of new refineries has proved to be poor, though the past failure has now been corrected. Natural gas could be an important instrument in supplying the shortfall of the middle distillates, but its impact is not likely to be felt until the late 1970s or early 1980s. A change in the price structure of oil products could serve to reduce the demand for gas oil considerably, although it is not certain whether it will be politically possible to raise kerosene prices.

In general, there seems to have been no coherent and detailed policy to deal with this important problem. For instance, there are no plans to deal with the excessive quantities of gasoline and fuel oil, no details of the future plans regarding new refineries, and no date has yet been set for laying the gas pipelines in the cities. Unless NIOC undertakes to implement some of the above alternatives immediately, under a coordinated set of policies, the domestic supply may be seriously affected in the very near future.

NOTES

1. Except for aviation fuel, which is exported to Afghanistan.
2. The author was informed by NIOC officials that the company would find it politically difficult to explain the need for importing oil products to a major oil-producing country.
3. The Russians constructed the Esfahan Still Mill Complex in partial payment for the natural gas.

CHAPTER

12

PRICES, COSTS, AND PROFITS

The purpose of this chapter is to investigate the various financial aspects of NIOC and to show by historical analysis the constraints, difficulties, policies, and achievements of the Iranian oil industry. The difficulty in such an analysis is to show the compromise between the financial aspirations of a company on a commercial basis and the wishes of the central government. This chapter tries to distinguish between such conflicts of interest and an evaluation of the rationality of various policies will be made on an economic basis.

PRICES OF OIL PRODUCTS IN IRAN

The NIOC Constitution obliges the Company to submit for approval to the cabinet any proposed changes in prices of the four main products--gasoline, kerosene, gas oil, and fuel oil.[1] In effect, the commercial right of the Company to determine the prices of products which constitute over 80 percent of its total sales revenues has been taken away by the government.

Since the nationalization of the Iranian oil industry, and the creation of NIOC, governments have frequently pointed out that they wish the oil product prices to be (a) uniform in the whole country, and (b) kept as low as possible, to provide a cheap source of energy for the industrialization of the country.

With regard to petroleum product prices, three major events have taken place since 1957. First, a switch to uniform prices for all products for the entire country was completed in 1964.[2] Second, in the process of equalizing the fuel oil prices, additional fuel oil taxes were levied in November 1964, while NIOC's after-tax realization was left unchanged.[3] Third, there was an experiment with

taxes on gasoline and kerosene in December 1964, which resulted in
an unprecedented public reaction, and the tax increase was almost
entirely rescinded within a few months. From 1965 onward the prices
of the four main products remained unchanged.

Although the Iranian economy is largely dependent on petroleum
products which constitute 70 percent of the country's energy requirements, any change in the prices of these products has proved to be
greatly resisted by the public. Petroleum has played a dominant role
in the historical development of Iranian politics and economics, and
as such it is very much under public scrutiny. There seems to be a
feeling among the people that at a time when Iran has established her
sovereignty over oil resources, the government should ensure that
these products are available to Iranians at the lowest price. Indeed,
the price increases of 1964 resulted in a fall in the absolute level of
demand for gasoline and a decline in the growth rate of demand for
kerosene. While it is generally agreed that the price elasticity of
demand is very weak with respect to petroleum products, due to the
absence of substitutes, it is clear that there is a political elasticity
of demand with respect to petroleum product prices.[4] Table 12.1
shows the changes in the prices of the main products since 1957.

TABLE 12.1

Prices of Petroleum Products in Iran
(rials per cubic meter)

	Gasoline	Kerosene	Gas Oil	Fuel Oil
1957–59	4,500	2,300	1,750	820
1960–62	5,000	2,300	2,250	820
1963	5,000	2,500	2,400	900
December 1964	10,000	3,500	2,400	1,200
1965	6,000	2,500	2,400	1,200
1974	6,000	2,500	2,400	1,200

Note: (a) The above prices include taxes, (b) the demand for
gasoline increased by over 5 percent in 1964, but declined by 3.4
percent in 1965. The demand for kerosene increased by 14.2 percent in 1964 and by only 2.3 percent in 1965. (c) The above prices
do not reflect the prices of "packed products."

Source: The Statistical Office of the National Iranian Oil
Company's Distribution Department.

Petroleum product prices are not only important because of the way they affect the revenues of NIOC, but also because they play an important role in shaping the pattern of demand for various products. Iran suffers from an acute imbalance between the pattern of supply and the structure of demand. This problem was discussed in detail in Chapter 11, but briefly we can say that the existing refinery capacity of the Iranian oil industry has been struggling to keep up with the large annual increase in the demand for middle distillates. To cope with the so-called "middle-distillate problem" NIOC has been, and is, building complex refineries[5] which maximize the yield of the middle-distillate output, and has lifted higher and higher proportions of kerosene and gas oil from the Abadan export refinery. These liftings from the Abadan refinery have resulted in a loss of potential exports of middle distillates with relatively higher export prices.

Thus, the relative prices of petroleum products in Iran are important insofar as they affect the pattern of demand for these products. A change in the structure of prices may help to balance the supply and demand, and therefore save the cost of building more complex refineries and the unbalanced liftings from Abadan.

DOES NIOC HAVE A PRICE POLICY?

In this section we will attempt to determine whether Iran has had a domestic oil price policy at all, and if so, on what basis it was formulated. Three possible policies will be considered: (i) pricing on the basis of export prices, (ii) cost-plus pricing, and (iii) pricing along the European pattern.

Pricing Based on Export Prices

An important approach to the problem of pricing would be to consider the economic opportunity cost of additional volumes of products demanded by the domestic economy. The present refinery capacity is given and for all practical purposes there is no choice but to use the NIOC-owned facilities to the utmost. This will satisfy a certain demand which might be called base demand. For volumes exceeding the base demand there are two alternatives: liftings from Abadan, or the building of a new domestic refinery capacity. The latter can range from topping units all the way to the most complex refineries.

For the Abadan refinery liftings, the relative values of products are determined by their export prices, which are reflected fairly accurately in the posted export prices for bulk cargo at Bandar Mashahr. Table 12.2 compares the posted prices of Iranian exports with the domestic product prices.

TABLE 12.2

Domestic and Export Prices of Iranian Oil Products, 1975[a]
(dollars per barrel)

Product	Export Price 1971[b]	Export Price 1975[b]	Domestic Price[c]	Domestic Price as a Percent of Export in 1975
Gasoline	3.83	15.83	14.11	89
Kerosene	3.95	14.70	5.90	40
Gas/diesel oil	1.90	13.23	5.60	42
Fuel oil	1.62	10.00	2.80	28

[a]All prices include taxes.

[b]Derived from the posted prices for BP ex-Bandar Mahshar, published in the Petroleum Economist, January 1971 and March 1975.

[c]Derived from Table 12.1.

Source: Compiled by the author.

We can see from the above table that domestic gasoline prices are 89 percent of the export prices, while kerosene prices are 40 percent of the corresponding export prices.

The above domestic and export prices are not strictly comparable, because the former exclude transport and marketing costs as well as local taxes, while the latter is a posted price and does not reflect the Iranian receipts. Nevertheless, one important point emerges: there seems to be no relationship between domestic prices and export prices--that is, there is no oil price policy in Iran based on the opportunity cost of exports.

Pricing Along the European Pattern

A study of the relative prices of petroleum products in Western Europe, where oil companies compete to some extent with each other, may yield some lessons for the Iranian oil industry. The most immediate problem in such a price comparison would be that of crude-cost differentials. In 1968 crude cost in Iran was about $0.12 per barrel, while the corresponding cost of imported crude in Europe was about $2.20 per barrel. To avoid this differential, 1.3 cents per liter of crude, and 1.4 cents per liter of product,is added to Iranian costs.[6]

The six European countries under study are the United Kingdom, France, Holland, Italy, West Germany, and Sweden. The products under study are gasoline, kerosene, automotive gas oil, gas oil for heating purposes, and fuel oil. Value-added tax and excise tax are included where appropriate.

TABLE 12.3

Average Prices, Average Tax, and Average Ex-Tax Prices for
Oil Products in Europe Compared with Iran, 1968
(cents per liter)

	Selling Price		Tax and Duties		Excise Tax Prices	
	Europe	Iran	Europe	Iran	Europe	Iran
Gasoline	16.8	9.4	11.9	5.2	4.9	4.2
		(8.0)				(2.8)
Kerosene	7.8	4.7	1.7	1.1	6.1	3.6
		(3.3)				(2.2
Automotive	11.0	4.6	7.9	0.9	3.1	3.7
gas oil		(3.2)				(2.3)
Heating	3.7	4.6	0.8	0.9	2.9	3.7
gas oil		(3.2)				(2.3)
Fuel oil	2.2	1.91	0.5	0.26	1.7	1.65
		(1.51)				(1.25)

Notes: (a) Figures in parentheses indicate the prices excluding crude cost differential. (b) Fuel oil of 275 viscosity for Iran and 800-1500 viscosity for Europe. (c) Although the European data are out of date because of the increases in cost of crude and excise tax in recent years, nevertheless the conclusions will remain unchanged.

Source: "Product Prices Abroad," unpublished NIOC-SRI report (1970).

TABLE 12.4

Iranian Price Structure for the Four Main Products Compared
with Average of Six European Countries, 1968
(Iran as a Percentage of European Average)

Products	Selling Price	Taxes and Duties	Excise Tax Prices
Including crude differential			
Gasoline (regular)	56	4	86
Kerosene	60	65	59
Gas oil (automotive)	42	11	119
Gas oil (heating and cooking)	124	120	128
Fuel oil (medium)	87	52	97
Fuel oil (light)	56	--	--
Excluding crude differential			
Gasoline (regular)	48	44	57
Kerosene	42	65	36
Gas oil (automotive)	29	11	71
Gas oil (heating and cooking)	86	120	79
Fuel oil (medium)	69	52	71
Fuel oil (light)	44	--	--

Source: "Product Prices Abroad," unpublished NIOC-SRI
report (1970).

Tables 12.3 and 12.4 are particularly informative, as they
provide a basis for comparison between the Iranian and European
price structure. With regard to "regular" gasoline, we can see
that the Iranian prices are about half the average pump price in the
six European countries. This is almost entirely due to lower taxes
in Iran (5.2 cents per liter versus 11.9 cents per liter European
average). NIOC's unit sales revenue, with ex-tax, compares to the
revenue of the oil companies in the United Kingdom, the Netherlands, and Italy, and amounts to about 86 percent of the European
average. This is interesting as it shows that the oil companies in
Europe are not charging excessively high prices, as many would
assume. Kerosene shows a different pattern. Including crude differentials, NIOC sales revenue (ex-tax) is 59 percent of the European average, while the selling price is 60 percent, with taxes at a

comparative level of 65 percent. However, this comparison may not be very useful because kerosene plays a minor role in Europe and therefore prices may not have much significance.

The analysis of gas oil prices shows a marked price difference in Europe according to end use, whereas in Iran the product is marketed at a uniform price. Automotive gas oil is sold at a comparatively low consumer price in Iran, even including crude differential (42 percent of the European average). On the other hand, NIOC's sales revenues, ex-tax, are higher than the European average by almost 20 percent. The reason for this is the comparatively very low tax, which in this case is a full 7 cents per liter below the European average. In this European sample, only the Netherlands has a tax burden more comparable with the one in Iran. But if gas oil for space heating is considered, the picture is entirely reversed. Including crude differential, the Iranian selling price is higher than in any of the European countries (124 percent of the European average). The taxes too, are higher in Iran than in the United Kingdom, France, West Germany, and Sweden, and amount to a startling 120 percent of the European average. As a result, the selling price (ex-tax) in Iran is quite high (128 percent of the European average).[7] A comparison for fuel oil is more difficult, because of the unusually low viscosity of fuel oil sold in Iran. In general Table 12.4 can be interpreted as follows.

Not surprisingly, the Iranian consumer actually pays considerably less for petroleum products than his European counterpart, with a low of 29 percent for automotive gas oil and a high of 86 percent for gas oil used for heating purposes. Interestingly enough, kerosene, which is the heating and cooking fuel for the vast majority of the Iranian population, is comparatively more expensive than automotive gas oil (with 42 percent of the European average) and much cheaper than gas oil used for heating.

The Iranian government collects considerably less taxes and duties than the European governments, on average, except on gas oil for heating purposes. It is interesting to note that the government levies on kerosene are comparatively higher than on other products. Also, thanks to the crude-cost differential, NIOC realizes fairly high sales revenues ex-taxes ranging from 59 percent for kerosene to 119-128 percent for gas oil.

From these observations, one can conclude that, compared with our European model, the Iranian government is in effect subsidizing automotive gas oil, while this business is very profitable for NIOC. Conversely, NIOC has a low realization on kerosene, while the government raises comparatively high taxes on this fuel. On balance, looking at the consumer prices, we can conclude that automotive gas oil is cheaper compared with other products. The prices for gasoline, kerosene, and light fuel oil fall within a surprisingly narrow range (42 percent to 48 percent of the European average). In order to increase the price of automotive gas oil to the lower end of this range (that is, 42 percent), its price should be increased from 3.2 cents to 4.6 cents per liter.

Cost-Plus Pricing

Cost-plus pricing is one of the most commonly used techniques of pricing. The basic idea behind cost-plus pricing is that the firm adds an increment to its average cost when deciding its sales prices. In this way the firm is expected to have profits proportional to the volume sold. We do not intend to discuss the advantages and disadvantages of such a system, but merely to determine whether a price policy based on costs has been in operation in the Iranian oil industry.

In such type of analysis the total cost of a barrel of refined oil must be considered and the incremental profit required must be added to this total cost figure. Table 12.5 provides the relevant data.

TABLE 12.5

Average Costs and Prices of Oil Products in Iran, 1957-73
(rials per cubic meter)

	Average Price	Average Cost	Profit Margin
1957	1,315	1,627	-312
1959	1,570	1,436	134
1961	1,582	1,559	23
1963	1,696	1,328	368
1965	1,725	1,280	445
1967	1,722	1,149	573
1969	1,716	1,109	607
1971	1,750	1,070	680
1973	1,860	1,280	580

Note: Average price excludes taxes but includes the price of "packed products."

Source: NIOC Distribution Department, Annual Reports 1966-73.

By plotting the data on Figure 12.1 we can get a picture of the movements of costs and prices. The profit margin has been increasing in most of the years, but not within a regular pattern. There is, however, no indication of a cost-plus pricing policy in Iran.

The above data relate to all the products taken together. There is no indication of the relationship between individual costs and prices. But can there be any measurement of individual costs where joint products are concerned? In the following section the problem of joint costs is discussed.

The Problem of Joint Costs

In the petroleum industry, joint costs enter into most of the downstream operations. We will, however, confine our discussion to the refining stage.

For a successful refining and marketing operation, total sales revenue must exceed total costs. Therefore, to assess the profitability of an oil company it is not necessary to know the individual costs of each product. Then why is there an interest in apportioning refining costs to various products? The reasons may include the accountants' wish to keep a tidy book of inventories, or the wish to decide the level of individual product prices by using the corresponding individual costs. Indeed, NIOC accountants, with the help of the Stanford Research Institute, have allocated the joint refining costs to various products. Such apportioning of costs may be done by allocating costs on the basis of demand or sales. In the case of NIOC, arbitrary ratios which were claimed to be based on "experience" were used for such allocations, and the author was not informed of the way these ratios were arrived at.

The danger with such cost allocation is that one may be led to draw conclusions from such data, forgetting the fact that such a division of costs is economically meaningless in the first place. Average costs can simply not be apportioned where joint products are concerned. The fallacy of allocating joint costs is pointed out in general economics text books, as well as more specialist writings. Professor Stigler writes:

> Such an allocation must be arbitrary, for there
> is no one basis of allocation that is more per-
> suasive than others. Indeed, any allocation of
> common costs to one product is irrational, if it
> affects the amount of product produced, for the
> firm should produce the product if its price is at
> least equal to its minimum marginal cost. [8]

FIGURE 12.1

Average Prices and Average Costs
in the Iranian Oil Industry, 1957–73

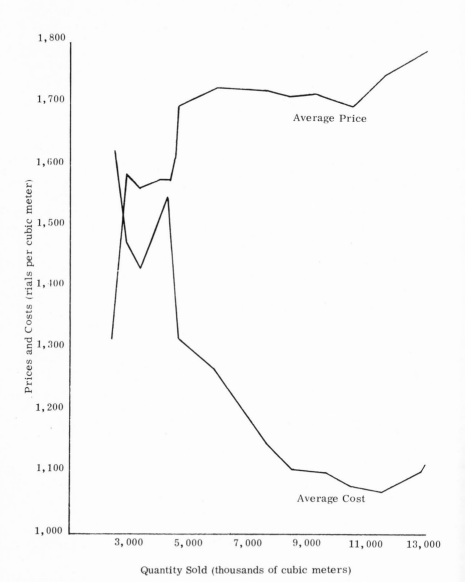

Note: Prices exclude taxes.

Insofar as oil is concerned, Porter writes:

> Accounting for refining operations begins with
> the receipt of crude oil and ends with the deliv-
> ery of the finished products. Between these two
> points there is a complex array of costs, includ-
> ing investment in materials and supplies, invest-
> ment in plant and equipment and operating costs.
> It is usually not difficult to determine which of
> these costs should remain in investment and
> which should be charged to operations for a par-
> ticular period. But it is virtually impossible
> to attribute particular operating costs to par-
> ticular products in anything more than an arbi-
> trary fashion.[9]

A very important distinction has to be made at this stage.
While it has become clear that total costs cannot be allocated to
joint products on an average cost basis, it is economically correct
to decide prices on the basis of marginal cost, even where joint
products are concerned. But before we consider such a possibility
in the case of the petroleum industry, let us look at two practical
attempts to solve the problem of joint costs.

A. A. Walters[10] deals with the problem of joint costs in the
transport industry. He considers the theory that marginal costs of
each product can be calculated by holding others constant while we
increase the production of one product by one unit. To find the most
profitable product mix we need to equate marginal cost and marginal
revenue for each product. There is no "allocation of joint costs" as
such, involved directly in this procedure; it has all been taken into
account in calculating the marginal costs. Walters introduces the
concept of "marginal expected costs" and sets up a mathematical
model for his problem. Although Walters argues that under certain
conditions "the allocation of accountants and traffic costing officers
may often have a rational basis," he observes that the round-trip
vehicle costs determined by his model vary greatly from those of
the accountants' allocations.

R. Weil, Jr. begins his article with an attack on "traditional
economists." He writes: "I suggest that management scientists
ought not to be led astray by the economists' traditional arguments
on allocating joint costs and that meaningful answers to reasonable
questions can be provided."[11] However, Weil's second paragraph
goes on to say: "I show the manager who wants an opportunity cost
or an accountant who wants a cost for inventory allocation how to determine these numbers, but I wish to emphasize that calling these numbers costs does not change the fact that they are _marginal_ revenues."

Both of the above articles provide practical methods of dealing with the joint costs problem. But it must be remembered that both approaches use the marginal concept.

Just how closely do all these arguments relate to joint refining costs? The basic idea is the same. All the products are produced jointly, whether some of these are wanted or not. At the same time, it is not possible to increase the output of one produce and hold the next constant for a given refinery pattern. Let us assume that furnace oil is considered undesirable by the refining company, but the production of a certain amount of this product is unavoidable. In this case no cost has been incurred to produce furnace oil, since it was not desired by the refining company, and to attribute a part of the total costs of refining--equal to its proportion in the total volume of oil production, or to its proportion in total sales--would be economically meaningless.

But how can we calculate the marginal cost of any one product of a refinery? This is possible because there are certain installations and equipment which may be added to a refinery to increase the yield of certain products (Table 12. 6).

TABLE 12. 6

Percentage Yield of Various Oil Products from the Same
Kuwait Crude by Applying Different
Refining Techniques

Methods	Light Distillates	Middle Distillates	Heavy Ends	Refinery Loss
Distillation and platforming	11	28	56	5
Plus visbreaking	13	38	42	7
Plus catalytic cracking	21	36	35	8
Plus propane deasphalting	39	27	28	10
Plus hydro-cracking	15	53	21	11

Source: Shell International Oil Company, "International Oil Prices," 1964.

As we can see from Table 12.6, any addition of a unit to a refinery can affect the output pattern. Thus if we intend to increase the output of light distillates by installing a catalytic cracking unit, the marginal cost of the increased output of light distillates would be roughly equal to (a) the cost of installation (including the future discounted cost of maintenance and operation) of the cracking unit, and (b) the discounted present value of the loss in earnings over the whole life of the cracking unit, because of the fall in the proportion of the other oil products. The marginal cost estimated in this way will have an important effect on the decision-making process. Unless the additional earnings over the whole life of the cracking unit, resulting from the increase in the proportion of gasoline, net of item (b), and discounted at least equals item (a), it would not be economically worthwhile to install the cracking unit.

Let us consider the special case of the Tehran refinery. Because of the need to increase the yield of the middle distillates, an Isomax unit has been installed there. It is estimated that a complex refinery (which includes special purpose units) costs $49 million more than a refinery of the same size. It produces 18,600 b/d of additional disillate products at the expense of 22,100 b/d of fuel oil. Also, operation cost in a complex refinery is about 22.5 cents per barrel higher than those in a simple refinery. On the basis of these figures, the Stanford Research Institute concluded that the marginal benefits were less than the marginal cost, and that NIOC may well be making a net loss on its sales of middle distillates.[12]

Despite the importance of marginal costs in the decision-making process of an oil company, it appears that NIOC has not formed its pricing policy on such a basis. The marginal cost of the special installations added to the Tehran refinery to increase the yield of middle distillates was in no way reflected in the prices charged for these products--indeed, there was no change in the prices at all. It is, however, important to note that NIOC has been unable to change its prices without the consent of the Cabinet. At the same time the company was obliged to produce middle distillates in accordance with the demand. Thus, regardless of the marginal costs involved, the company was forced to install the necessary equipment when required.

Alternative Criteria for Pricing Policy

In the following, four alternative criteria for pricing policy will be discussed.

(a) Maximizing net revenues from a barrel of refined oil may well be the objective of a price policy for an oil company. Prices may be determined by using the elasticity of demand (and the cross-elasticity of demand for substitutes) for various products. In this way a certain level of price may be (mathematically) decided for each product which will maximize net revenues from a barrel of refined oil. One important point, however, has to be made at this stage. It is not necessary for the price of every barrel of a certain product to exceed the cost of crude oil. The price of heavy fuel oil (HFO) provides an example. HFO is a simple combustible, which is worth no more than any other source of heat, allowance being made for handling costs a little lower than coal and a little higher than gas. All other refined products have high value uses with no near substitutes. Thus the HFO price cannot go much higher than crude, for if it did consumers would burn the entire crude. But HFO prices can go below crude, or even to zero and below (when it involves a disposal cost).

In the words of Professor Adelman,

> . . . the prices of other products (other than HFO) cannot go below the price of crude, for then it would not pay to refine them out. They would only have fuel value. Therefore, the basic rule is: the more are the prices of other products above crude oil, the more is heavy fuel oil below it. Conversely, the nearer the prices of other products are to their lower limit (the price of crude oil) the closer is the heavy fuel oil to its upper limit, which is also the price of crude oil. At this point there is no longer any refining.[13]

Adelman goes on to say that the competitive opportunity cost of a barrel of crude oil must exceed that of HFO, because crude can profitably be made into more valuable products, while fuel oil cannot. To charge an equal or higher price for fuel oil than for crude is to discriminate in price against fuel oil buyers.[14]

(b) Encouraging industrial use may be a criterion for pricing policy. The government may wish to encourage the expansion of petroleum based industries and thus, by keeping industrial product prices steady in the face of domestic and external inflation or by discriminating between the domestic and industrial prices, the government will in effect subsidize the industrial users. In the case of import substitution, the subsidization of an essential industrial input, such as petroleum products, will have the effect of a high rate of effective tariff protection. Such protection of domestic industries may be considered in the case of infant industries. In Iran, the steady prices of fuel oil and gas oil during the 1960s led to an increasing level of industrial activity (see Chapter 10).

(c) Subsidizing low-income budgets may be feasible. The government may wish to use the prices of oil products to subsidize lower-income groups. This subsidization may take place through lower prices of kerosene, and possibly gasoline. Our analysis of the Urban Household Budget Survey of Iran showed the importance of the oil product expenditure in the lower income groups (see Chapter 10).

(d) To correct the product imbalance, a pricing policy may be devised in such a way that relative prices will affect the demand pattern and thus reduce the imbalance between supply of and demand for oil products (see Summary and Conclusion).

One may use the following approach to discuss the above alternative price policies. The maximization of the net revenue from a barrel of products may be treated as the objective function of the company. The firm will attempt to maximize its objective function in the face of the prevailing constraints such as demand constraint, technical constraint, and cost constraint. In this way other objectives, such as encouraging industrial use, subsidization of low-income budget, and correcting the product imbalance, may all be treated as subsets of the original objective functions. That is, these other objectives may be treated as constraints in maximization of the objective functions.

In the case of Iran, various criteria have been used to see whether NIOC has followed a certain price policy. It appears that none of these criteria conform to the Iranian prices. Moreover, the author's discussion with NIOC officials confirmed the point that NIOC has not followed a particular price policy based on any economic criterion. However, the government freeze on the prices of the four main products since 1965 has resulted in subsidization, particularly in the case of kerosene. Although any kind of subsidization can be treated as a price policy, it is fair to say that such subsidization was a side effect of the government freeze, rather than the result of a deliberate economic policy.

COSTS IN THE IRANIAN OIL INDUSTRY

Costs constitute the largest element in the prices of petroleum products in Iran, although it is true to say that prices are independent of costs. Since 1962, costs have contributed to around half of the price, as shown in Table 12.7.

TABLE 12.7

Components of Average Price in Selected Years
(price = 100)

Year	Taxes and Duties	Costs	Profit Contribution
1962	34.2	57.2	8.6
1965	39.0	45.3	15.7
1968	34.5	42.5	23.0
1971	34.7	40.0	25.3
1973	34.2	45.5	20.3

Source: Computed from the data in the Annual Reports of the
Distribution Department.

Costs are divided into three major categories: transport
costs, which includes the delivery of petroleum products from refineries by pipeline, tankers, railways, and barges; product costs,
which include costs of crude, cost of transporting crude to the refineries, and actual refining costs; finally, distribution costs, which
include sales expenses, divisional expenses, marketing and management overheads, commissions, depreciation, and the cost of containers.[15]
During the 1962-72 period, transport costs were by far the
largest cost component, accounting for about 45 percent of the total
costs in this period. Product costs have generally declined over this
period from 32 percent of the total in 1962 to 28 percent in 1972.
The reduction in cost of products was the result of the Consortium's
efficiency in reducing production costs and increasing labor-saving
installations at the refineries. Distribution costs constitute the
third largest category, or about 25 percent of the total in 1972. Generally speaking, transport and product costs are more likely to fall
per unit sales than the distribution costs. This is so because a large
portion of these costs are fixed and do not increase in direct proportion to the volume of sales; but distribution costs are heavily biased
toward current costs, particularly commissions, which have soared
in recent years.
It is important to note that the problem of joint costs (discussed
earlier) enters into any calculation of costs for individual products.
There is no logical way whereby one can attribute different costs to
different products. Thus no attempt will be made to estimate individual product costs.

PROFITABILITY OF DOMESTIC SALES

Profits in the Iranian oil industry are a function of various costs (transport, distribution, and refining), taxes and duties, and as such they are governed by the changes in these variables. Profits on product sales have risen from a low of 8.6 percent in 1962 to 20.3 percent in 1973 of the average price (see Table 12.7). The reason behind this sharp increase has been the large scale of NIOC's domestic operations and the resultant decline in various cost items.

To assign profits to individual products will, of course, involve the arbitrary allocation of the total refining costs to individual products, and as such the profit figure for each product will be arbitrary. Therefore the only economically meaningful figure is that of total profits. Table 12.8 shows the profit margin per cubic meter of domestic sales.

TABLE 12.8

Profitability of Domestic Sales, 1957-73
(rials per cubic meter)

Year	Profit Margin
1957	-312
1959	134
1961	23
1963	368
1965	445
1967	573
1969	607
1971	680
1973	580

Source: Annual Reports of the Distribution Department.

NIOC'S FINANCIAL STATEMENTS

NIOC publishes its financial statements every year, although they are not widely distributed. The published data show the consolidated accounts of the company which includes all its direct activities both at home and abroad. As of 1973, the assets of the Consortium are reflected in NIOC's balance sheets as a result of the 1973 Sales and Purchase Agreement.

The Distribution Department of NIOC (DD) also publishes an annual income statement for the domestic sales of oil products in Iran which shows the profits earned by DD and gives an indication of the contribution of domestic profits to NIOC's expansion in recent years.

NIOC acts as a tax-collecting agency for the government. All the oil payments as well as royalties, bonuses, and balancing margin are collected by NIOC and passed over to the Ministry of Finance and Economics. Also, NIOC collects excise taxes on products, and taxes payable on the salaries of the employees and contractors for the government. The company itself is required to pay an annual dividend despite its meager profits, while its profits are taxed at the rate of 49.47 percent--like any other private company in Iran.

NIOC's sources of finance are (a) the net trading profit on the sale of oil products in Iran (after tax), (b) depreciation allowance, (c) funds received from the Plan Organization (now Plan and Budget Organization), and (d) loans from government and banks, including occasional credits from the World Bank. In 1966 the largest element of NIOC's receipts was loans, which were nearly twice as large as the company's cash flow (net after tax profits plus depreciation). From 1966 onward the loans declined in relative importance while the contribution of government through Plan Organization greatly increased. In 1969, the latter funds were 3.5 times NIOC's cash flow and 5.5 times the company's after-tax trading profits. The corresponding ratios in 1973 were 2.0 and 4.7.

NIOC is entitled to keep 2 percent of the Stated Payments (royalties) payable to the Iranian government by the foreign operators, for its general reserves. This figure was reduced to 1 percent under the 1974 Petroleum Act and NIOC Constitution (Article 48). This was decided in view of the increased receipts of the Stated Payments from 12.5 percent of the posted price in 1973 to 16.67 percent in 1974 under a general OPEC agreement.

Appendix Tables A.4 and A.5 show the balance sheet and the income statement of NIOC. The following observations may be made.

(a) Plant, property and equipment, net of depreciation increased from 12.7 billion rials ($166 million) in 1966 to 60.4 billion rials ($895 million) in 1973--a fivefold growth in NIOC's fixed assets in seven years. This represents the scale of investment in the Tehran, Shiraz, Madras, and Sasolberg refineries as well as investment in subsidiary and affiliated companies. The latter item in 1973 alone amounted to $365 million. The increase in fixed assets also reflects increase in capacity and number of pipelines together with other additions to the company's buildings, machinery, and vehicles. (The possibility of a revaluation of the assets cannot be ruled out.)

(b) Net assets (net worth) of NIOC increased from 22.8 billion rials ($298 million) in 1966 to 114.5 billion rials ($1,700 million) in 1973. The latter figure includes the transfer of the Consortium's assets to NIOC.

(c) The net trading profit of NIOC from domestic sales rose from 635 million rials ($8.5 million) to 4.8 billion rials ($71 million), representing an eightfold growth in profits in the 1966-73 period. This was achieved because of the large increase in net sales--$158 million in 1966 to $414 million in 1973--and a general reduction in unit costs.

(d) Cash flow of the company increased from 1.5 billion rials ($20 million) in 1966 to 5.8 billion rials ($86 million) in 1973, providing NIOC with greater, though insufficient, scope for expansion.

Table 12.9 shows the trading profits of NIOC as well as the profits accrued from the Distribution Department.

TABLE 12.9

Profits in the National Iranian Oil Company, 1966-73
(millions of rials)

	1966	1969	1972	1973
Sales in millions of cubic meters	6.7	9.6	12.8	15.4
Net trading profit of NIOC	653	2,154	5,155	4,805
Net profits of the Distribution Department	3,300	5,800	8,600	8,700
Net profits of NIOC after 50 percent tax	326	1,077	2,577	2,402
Cash flow	1,548	3,857	5,527	5,827

Source: National Iranian Oil Company.

The above table shows that the profits accrued from the Distribution Department have been substantially above those shown by NIOC. This can be explained by the fact that DD bears very little of the depreciation expenses--that is, the depreciation taken into account by DD is only on the facilities used by the Department and excludes depreciation on all other NIOC facilities. Also, the cost of research and development as well as overhead costs are not reflected in DD's

total costs. Whether the Distribution Department can become a
profitable entity on its own cannot be determined from the available
data.

Two very important points emerge from our analysis. First,
if we compare the growth of the net worth of the company to its
after-tax profits or its cash flow, it becomes clear that NIOC's expansion has been financed by external sources. Second, the fact that
NIOC's share of the vast oil revenues accrued to Iran through exports
is negligible, and the fact that NIOC is subject to 50 percent income
tax, like all private companies in Iran, reflect the government's attitude of keeping a close grip on the company's activities. To be
sure, the government has been very generous in providing funds for
NIOC's expansion through the Plan Organization, but through this
method of financing NIOC is obliged to seek the approval of the planning authorities as well as the Ministry of Finance and Economics
for its major projects. In this way any policy of expansion at home
or abroad and any diversification plans which NIOC might have is
kept in line with the general economic policy of the government.

CROSS-SUBSIDIZATION IN THE IRANIAN OIL INDUSTRY

Cross-subsidization has, in the last few decades, been an integral part of policy in public enterprises. It is a direct government
intervention to provide a service for one group of people, which is
subsidized by another group. In the United Kingdom, cross-subsidization is practiced in railways and coal production.[16] In the case of
the Iranian oil industry, we have to deal with various types of cross-
subsidization, each with quite different political and economic implications.

In the British coal industry, cross-subsidization operates on
production costs, but in the case of the Iranian oil industry it operates on transport costs and takes various forms.

The Hidden Subsidy Principle

Oil products in Iran are offered at uniform delivered prices.
The uniformity of prices all over the country, whether the centers
of consumption are close, or far from the refineries, creates the
problem of "hidden subsidy," due to differences in transport costs.
The problem can be set out as follows:

Components of price (P) = Profit (Pr) + Transport cost (Tr)

+ Distribution cost (d) + Cost of crude (C) + Cost of refining

(R) + Taxes (T)

thus

and

$$P1 = Pr1 + Tr1 + D1 + C1 + R1 + T1 \text{ in Center 1}$$

$$P2 = Pr2 + Tr2 + D2 + C2 + R2 + T2 \text{ in Center 2}$$

Since prices are uniform,

$$Pr1 + Tr1 + D1 + C1 + R1 + T1 =$$

$$Pr2 + Tr2 + D2 + C2 + R2 + T2$$

For the sake of convenience, we assume D, C, and R are equal in all regions, so Pr1 + Tr1 + T1 = Pr2 + T2. This means that in a simple model, profits are a function of transport costs and taxes. Since taxes are uniform all over the country for all products (except for fuel oil) we can safely assume that profits are a function of transport costs. This means that the regional profits are dependent on the distance from the refineries and centers of production.

Let us consider the Tehran refinery supplying products to five centers:

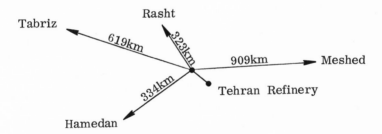

Taking an average transport cost of 0.57 rials per ton-kilometer, we can roughly calculate the transport costs to Tehran from:

1. Tabriz   0.57 X 619   = 352.8 rials per ton
2. Meshed   0.57 X 909   = 518.1 rials per ton
3. Hamedan   0.57 X 334   = 190.4 rials per ton
4. Rasht   0.57 X 323   = 184.1 rials per ton

We can now see the amount of subsidy per ton by considering the transport costs of each center of consumption as compared to Tehran. For example, Tabriz receives a transport cost differential subsidy of 353 rials per ton. This is equal to a subsidy of 304 rials per cubic meter or 2 cents per gallon. Although this amount of subsidy may not seem significant in U.S. terms, it is important in the

Iranian context. On average, one cubic meter of oil was sold for
1,700 rials net of taxes in 1969 (see Table 12.5); thus Tabriz received a subsidy of 18 percent compared to Tehran. As Tabriz consumed over 575,000 cubic meters of oil products in 1969, this means
that the subsidy to Tabriz amounted to 175 million rials, or $2.6 mil-
lion. Considering the population of Tabriz of 810,000 (1966 Census),
the subsidy is equivalent to $3.20 per person per year.

The measurement of the transport subsidy based on distance is
a relative concept. Tabriz receives a large subsidy compared to
Tehran, but would receive a smaller subsidy compared to Rasht,
which is closer to Tehran.

Transport cost subsidies become even more significant when
we consider the transport cost differential from Abadan compared
with, say, Tabriz. The cost of finished products in Abadan was 260
rials per cubic meter in 1968, compared with 648 rials at the Tehran
refinery. This was because crude oil had to be transported to the
Tehran refinery from the southern oilfields. When we compare
Tabriz to Abadan, we must enter into our calculations the product
cost differentials, which would magnify the amount of the subsidy.
The above analysis shows that some centers are enjoying a considerable subsidy at the expense of those who live near the centers of
production and refining. That is, prices charged are not based on
cost consideration.

Given the relevant data, one could construct a comprehensive
econometric model, showing transport and product cost subsidies
received by each center and hence the economic prices chargeable
to every region. Such data, however, were not available to the
author, but even without it one can roughly come to some broad conclusions. The most immediate conclusion would be that pretax
prices of the centers of consumption near a refinery would have to
be reduced by about one-fifth, since transport costs constituted 23
percent of the average pretax prices in 1969.

If Tehran is not to pay any transport costs and other regions
have to pay in full for theirs, then we may well find ourselves in a
situation where some regions will have to pay up to three times more
for their products. But will people be able to afford these so-called
economic prices? And is the elimination of the transport cost subsidy a practical proposal? In answering these questions we would
have to deal with three basic points. First, the political implications
of price increases referred to earlier. Second, the uniformity of
prices with its "noneconomic" basis has one great virtue: it is ad-
ministratively easy to operate and enforce. Third, one cannot even
be sure that such a proposal is economically justifiable. There is
no reason why the inhabitants of Tehran, who enjoy the highest standard of living in the country, should benefit from the lower cost of

petroleum products just because they happen to live near a refinery.
Traditionally the northern part of Iran has been the most prosperous
region, followed by the central region, while the southern regions
have been the least prosperous areas. Zahedan and Kerman in the
southeast, being among the poorest, accounted for only 2.9 percent
of the total consumption of oil products in 1969. There may be a
case for making available at lower prices to the underdeveloped regions some of the oil products such as kerosene and fuel oil. If one
were to adopt a purely economic approach, the poor inhabitants of
Zahedan, who "happen" to live about 1,500 kilometers away from the
Abadan refinery, would suffer greatly. There is also a great deal of
economic dualism within every region. For example, Esfahan is a
relatively prosperous city, but the villages and small towns around
it are poor. The Esfahanis may be able to afford high oil prices
which are too high for the other towns and villages in the Esfahan region, and of course it is impossible to have different prices for every
town and village in the country.

If the government is to avoid lopsided development (which is,
incidentally, a characteristic of capitalism in the developing regions),
it will have to base its prices on different grounds than the transport
cost differentials. It will have to be based at a macro-level on the
broad objective of balanced growth for all regions. The power of
government taxation can play a great part in these "balances," but
this power has not been used so far, with regard to oil products.

FUEL OIL TAXES AND PRICES

Unlike other oil products, which are sold at uniform pretax
prices, NIOC sells fuel oil at different pretax prices. NIOC offers
four different prices: one for the city of Tehran, one for the Kermanshah region, one for the Khuzistan region, and one for all other
regions. Except for the Khuzistan region, which is allowed to have
cheap fuel oil, the government seems to have indiscriminately imposed taxes on NIOC's prices to bring all the prices to 1,200 rials
per cubic meter. This policy places the greatest burden on Kermanshah, one of the regions least able to afford the standardized price.
And given that Kermanshah's refinery makes the region self-sufficient,
it is hard to envisage a sound justification for the present tax rate of
110 percent. Unfortunately, it seems that not only has the government not used its taxation powers to help the underdeveloped regions
(except for Khuzistan), but, in some cases, has helped to aggravate
the situation.

TABLE 12.10

Fuel Oil Taxes and Prices
(rials per cubic meter)

Region	Pretax Price	Taxes	Sales Price	Tax as Percent of NIOC Price
City of Tehran	866.1	333.9	1,200.0	38.5
Khuzistan	598.2	151.8	750.0	25.4
Kermanshah	575.0	625.0	1,200.0	108.7
All other regions	1,133.9	66.1	1,200.0	5.8

Source: The Statistical Office of the National Iranian Oil Company's Distribution Department.

In general, a great deal of subsidization and cross-subsidization has taken place in the Iranian oil industry. However, there is no indication that subsidization has taken place as a result of a clear government policy. Indeed, it all seems to be the by-product of the government's rigid attitude to the uniformity of prices throughout the country.

SUMMARY AND CONCLUSION

This chapter brought together the effects of the various decisions taken by NIOC and the government on prices, costs, and profits.

Prices of the four main petroleum products have been frozen since 1965, and all the oil products are sold at uniform prices throughout the country. Any proposed changes in the prices of the four main products, which accounted for 83 percent of the total NIOC sales revenues in 1973, are subject to ratification by the Cabinet. In this way the government has effectively removed NIOC's price determining powers.

The government's case for price stabilization rests on three arguments:

(a) The infant-industry argument: The Iranian industries are at an early stage of development and must be subsidized through lower input prices. The lower price of inputs would amount to a high rate of effective tariff protection for the domestic industry.[17]

(b) The importance of petroleum expenditure in the average household budget: Our analysis of the role of oil product expenditure in the Household Budget Surveys in 1967 and 1969 showed that in the lower-income groups the expenditure on health, housing, education, and clothing was often exceeded by expenditure on petroleum products (see Chapter 10).

(c) The possibility of strong reaction to changes in prices, similar to the large demonstrations in 1964.

In the first place, the infant-industry argument is not applicable to Iran, because it is already an extremely protectionist country and there are a variety of tariff and quota tools which the government can use to protect domestic industry. Moreover, too much protection may lead to inefficiencies in the long run. Second, in a typical household budget, kerosene and gasoline constitute the principal expenditures on petroleum products, and thus there is no case for stabilizing other product prices. Finally, the public disorder in 1964 was caused by increases of 100 percent in gasoline, and 40 percent in kerosene prices.

If the price rises are gradual and spread over a longer period, it is doubtful whether there would be any strong public reaction. On the other hand, the public sensitivity is directed toward gasoline and kerosene prices, and not toward industrial fuels.

Given the structure of prices, this chapter attempted to investigate whether there has been any form of price policy in existence in Iran. Three criteria were used to establish whether the product prices bear any relationship to other economic variables: export prices, costs, and international prices. The conclusion was that there has never been a price policy in Iran with regard to oil product prices. The only guideline for prices seems to have been the government's pressure on NIOC to maintain stable prices.

Price Policy and Energy Policy

Iran has no declared energy policy. Whether there is an energy policy in the minds of the Cabinet Ministers is another issue. In the past two decades no energy policy was required because of the ability of the domestic refining capacity to supply the domestic demand. In the early 1970s the need for a long-term energy policy became apparent, particularly in view of the failure to take appropriate actions in the past. It is the author's belief that no energy policy in Iran could succeed without a corresponding price policy. A change in the relative position of prices is required to change the pattern of demand for middle distillates. For instance,

a change in the relative prices of automotive gas oil and gasoline can
go a long way in reducing the middle-distillate shortage. The price
of gasoline is 8 cents a liter, of which tax represents 65 percent,
while diesel oil has a price of 3.2 cents a liter with a tax component
of 28 percent--the difference encourages the use of diesel oil. Thus,
either gasoline tax must be reduced or diesel oil tax increased or
both. Another way in which the problem may be tackled is to dis-
courage the import of diesel-fueled automobiles and trucks through
quotas or tariffs. Alternatively, a two-tier price system may be
adopted with a refund or discount for the users of gasoline trucks.

In Europe, the automotive and heating gas oil have different
prices, while the corresponding prices in Iran are identical. Diesel
oil prices may be raised without any increase in heating gas oil. To
avoid a shock to the transport sector and a sudden increase in prices
of agriculture and other goods which require long-distance transport,
the increase may take place over a period of time, combined with
other measures referred to before.

The government may find it politically difficult to raise the
prices of kerosene, but in the next few years a gradual increase may
be possible with the increase of the disposable income of the lower-
income groups. In the areas where gas oil and fuel oil compete as
industrial fuels, a small increase in the price of fuel oil will help to
reduce the demand for gas oil and alleviate the middle-distillate
shortage. All such measures, involving the combination of fiscal
and pricing policies, must be undertaken as a package deal to avoid
the chances of disruptive economic side effects.

Our analysis of the profitability of the domestic industry
showed that a general improvement has taken place in the past
decade. However, NIOC's sources of finance have come under
more and more government control, with the result that the company
can do little without the approval of the financial and planning authorities. In short, the growth of NIOC has been mainly financed by non-
domestic activities.

Cross-subsidization is an integral part of many public enterprises all over the world. This is so because the government-owned
enterprises are not concerned with business objectives, but need to
take into account the whole range of social, political, and economic
considerations. The compromise which determines the order of
priorities of these objectives constitutes the basis of the public policy
for a government venture. In Iran the political consideration of uniformity of prices has long overruled the other objectives in the domestic oil industry. The uniformity of prices has led to a major
type of cross-subsidization: hidden subsidy due to transport cost
differential. It would be unreasonable to suggest that transport
costs should become chargeable in full to the consumers, as this

would mean that some consumers, living in centers of consumption far away from refineries, will have to pay several times more for their products, and many of these centers of consumption are economically underdeveloped.

If the government of Iran is concerned with a policy of balanced regional development, it can use the prices of petroleum products, particularly those of industrial fuels, to achieve its goals. The prices of these products will have to be lowered in the underdeveloped regions, relative to the prices of petroleum products in the more prosperous regions. But NIOC cannot be expected to do this on its own. Any changes in price will have to be coordinated with the Ministry of Finance and will require reshaping of the tax structure of the petroleum products. Up to now, these taxes have been used solely to equalize prices throughout the country.

NOTES

1. NIOC Constitution, Article 35(b).

2. Special price discounts are available for Khuzistan farmers (in the southern region), Ministry of Water and Power and the Iranian Railways, etc.

3. Fuel oil prices were fixed at 0.75 rials per liter for Khuzistan and 1.2 rials per liter for the rest of the country.

4. See Chapter 10.

5. The term "complex" is applied to plants with isomerization and other equipment, so that the yield of middle distillates is increased. In technical language:

- A simple refinery consists of crude topping, catalytic reforming, and kerosene unifining (similar to Kermanshah design).
- A complex refinery consists of crude topping, vacuum distillation, middle-distillate hydrocracking, catalytic reforming, and visbreaking (similar to Tehran and Shiraz design).

6. M. A. Adelman, The World Petroleum Market (Baltimore: Johns Hopkins Press, 1971), p. 190. According to Adelman, a barrel of refined oil products in Rotterdam in 1968 was valued at $2.53, while the refining cost was $0.33, giving a crude price of $2.20 per barrel. The 1.3 cents differential is calculated in the following manner: 220¢ - 12¢ = 212¢: 158.98 = 1.3¢. Assuming that the refining costs in Iran are 0.10 cents a liter higher than those in Europe, then a product differential of 1.4¢ a liter would emerge.

7. Note that in 1974 the residential and commercial market in Iran was estimated to have consumed 261,000 cubic meters of gas oil

for heating purposes, while the transport market consumed over 1.7 million cubic meters of automotive gas oil (diesel oil). Gas oil sold to both markets was at uniform prices. For details see Chapter 10.

8. G. J. Stigler, The Theory of Price (New York: Macmillan, 1966), pp. 162-65.

9. M. Porter, Petroleum Accounting Practices (New York: 1966), p. 418.

10. A. A. Walters, "The Allocation of Joint Costs with Demand as Probability Distribution," American Economic Review 50 (1960).

11. R. Weil Jr., "Allocating Joint Costs," American Economic Review 58 (1968).

12. Stanford Research Institute, "Long Term Future Energy Policy in Iran," Report to the Imperial Government of Iran 1969 (unpublished). Sources and methods of estimates were not disclosed to the author.

13. Adelman, op. cit., p. 177.

14. Although Adelman was concerned with wholesale prices in a competitive market, his analysis demonstrates that NIOC-SRI accountant's figures showing losses on the sale of fuel oil are economically meaningless.

15. For a detailed breakdown of the cost figures, see F. Fesharaki, "The Development of Iranian Oil Industry 1901-71," Ph.D. dissertation, University of Surrey, March 1974, and Stanford Research Institute, op. cit., Vol. 5.

16. For a survey of various discussions on cross-subsidization see R. Turvey, ed., Public Enterprises (London: Penguin Modern Economics Series, 1968), particularly W. G. Shepherd, "Cross-Subsidization in Coal," pp. 316-51.

17. For details of the Theory of Effective Tariff Protection, see I. Little, T. Scitovsky, and M. Scott, Industry and Trade in Some Developing Countries, O.E.C.D. Development Center (London: Oxford University Press, 1970).

APPENDIXES

TABLE A.1

Projected Overall Government Finances
During the Fifth Plan

	Billions of Rials	Billions of Dollars
Receipts		
Oil and gas	6,628.5	98.2
Direct taxes	547.0	8.1
Indirect taxes	668.0	9.9
Other receipts[a]	253.0	3.7
Foreign loans	150.0	2.2
Banking credits (net)	--	--
Sale of government bonds (net)	50.0	0.7
Total receipts	8,296.5	122.8
Payments from general revenue		
Current expenditures	3,393.3	50.2
General affairs	(452.8)	(6.7)
Defense affairs	(1,968.7)	(29.1)
Social affairs	(754.0)	(11.1)
Economic affairs	(217.8)	(3.2)
Fixed-capital formation	2,848.1	42.2
Repayment of principal of foreign loans	405.0	6.0
Other payments	905.0	13.4
Investment abroad	745.1	11.0
Total payments	8,296.5	122.8

[a]Includes 135 billion rials ($2 billion) revenue from public
sector investment in and loans to other countries.

Source: Fifth Development Plan of Iran (1963-68), p. 41.

TABLE A.2

Gross Domestic Product by Industrial Origin at (1959) Constant Prices, 1962–72
(billions of rials)

	1962	1963	1964	1965	1966	1967	1968	1969	1970	1971	1972
Agriculture	88.8	90.3	92.2	99.5	103.0	111.1	119.7	123.4	129.1	124.4	134.4
Oil	67.3	73.9	83.6	93.9	108.5	127.4	145.4	167.8	195.0	221.9	250.5
Manufacturing and mines	41.5	45.2	47.2	53.7	62.9	72.5	82.8	90.9	100.9	118.1	132.3
Construction	14.1	16.3	17.2	22.2	21.4	24.9	26.2	26.6	27.3	33.0	37.6
Water and power	2.2	4.2	4.8	5.7	7.7	8.9	10.8	13.4	16.2	19.5	22.8
Transportation and communication	30.0	30.8	31.7	33.3	34.3	35.6	37.5	40.0	44.7	46.7	53.8
Banking and insurance	7.5	8.1	9.4	10.6	11.9	14.0	18.4	22.7	28.9	32.1	34.4
Domestic trade	24.4	25.3	28.5	31.5	35.5	39.8	43.1	46.8	51.2	56.2	62.5
Ownership of dwellings	19.2	20.3	21.9	23.8	25.8	27.7	29.9	31.9	34.1	36.7	39.9
Public services	25.2	29.5	34.9	43.3	46.9	49.5	57.2	62.4	72.4	87.9	103.2
Private services	14.2	14.8	16.0	18.1	19.7	21.6	26.1	27.6	32.2	38.5	44.2
Gross Domestic Product at Factor Cost	334.3	358.7	387.4	435.6	477.6	533.0	597.1	654.5	732.0	815.0	915.6
Growth rate of GDP	6.7%	9.3%	8.0%	12.4%	9.6%	11.6%	12.0%	9.6%	11.8%	11.3%	12.3%
Net indirect taxes	20.8	21.8	24.4	26.9	30.8	34.4	37.5	40.3	43.4	47.6	50.6
Gross Domestic Product at Market Prices	355.1	380.5	411.8	462.5	508.4	567.4	634.6	694.8	775.4	862.6	966.2
Nonoil GDP at Factor Cost	267.0	284.8	303.8	341.7	369.1	405.6	451.7	486.7	537.0	593.1	655.1
Growth rate of nonoil GDP	4.7%	6.7%	6.7%	12.5%	8.0%	9.9%	11.4%	7.8%	10.3%	10.4%	12.1%
Growth rate of oil GDP	15.8%	9.8%	13.1%	12.3%	15.5%	17.4%	14.1%	15.4%	16.2%	13.8%	12.9%

Source: Bank Markazi Iran

TABLE A.3

Gross National Product in Selected Years (1972 Prices)

(billions of rials)

	1967-68	1972-73	1977-78	Average Annual Rate of Growth During the Fourth Plan Period (percent)	Average Annual Rate of Growth During the Fifth Plan Period (percent)
Consumption expenditure	540	898	2,168	10.7	19.3
Private sector	(442)	(645)	(1,322)	(7.9)	(15.4)
Public sector	(98)	(253)	(846)	(20.8)	(27.3)
Gross domestic fixed capital formation	151	287	1,052	13.7	29.7
Private sector	(77)	(141)	(319)	(12.9)	(17.7)
Public sector	(74)	(146)	(734)	(14.6)	(38.1)
Balance of payments, current account	-5	-20	465	--	--
Gross national product, at market prices	686	1,165	3,686	11.2	25.9
Population (millions)	26.5	31.0	35.9	3.0	2.9
Per capita GNP (rials)	25,894	37,522	102,665	7.7	22.3
Per capita GNP (U.S. dollars)	384	556	1,521	7.7	22.3

Source: The Fifth Development Plan of Iran, Plan and Budget Organization, Tehran, 1975, p. 27.

NIOC's Balance Sheet, 1966-73
(millions of rials)

	1966	1969	1972	1973*
Assets				
Fixed assets after depreciation	12,692	43,380	69,554	60,419
Movable assets after depreciation	--	509	1,046	11,772
Investments and loans	3,877	19,784	27,005	28,129
Current assets	11,779	18,640	34,767	73,773
Less: current liabilities	6,140	16,270	25,267	59,525
Excess of current assets over current				
liabilities	5,639	2,370	9,500	14,248
Net assets	22,208	66,043	107,105	114,568
Liabilities				
Share capital				
Authorized--10,000 shares of				
1,000,000 rials each	10,000	10,000	10,000	10,000
Issued and fully paid--5,000 shares				
of 1,000,000 rials each	5,000	5,000	5,000	5,000
Reserves				
Capital reserve	7,820	48,796	29,555	36,108
General reserve	491	--	3,821	5,596
	8,311	48,796	33,376	41,704
Funds received from Plan Organization	7,443	5,130	60,811	25,982
Long-term loans	1,454	7,117	7,918	7,301
Advance against future deliveries of oil	--	--	--	34,581
	22,208	66,043	107,105	114,568
Memorandum accounts				
Fixed assets in Consortium Agreement				
Area	15,500	19,260	24,118	--
Amoco-Iran Oil Company exploration				
expenses receivable	2,155	2,176	1,097	442
Treasury notes purchased against loan				
received from trading companies				
for Khuzistan development	1,254	1,277	1,444	1,594
Loan granted to Khuzistan industries	--	156	395	392
Guarantees received	--	--	696	988
	18,909	22,869	27,750	3,416
Less:				
Right to use and benefit from fixed assets				
by operating companies in Consortium				
Agreement Area	15,500	19,260	24,118	--
Iran Pan-American Oil Company				
exploration expenses payable	2,155	2,176	1,097	442
Loan received by government from trading				
companies for Khuzistan development	1,254	1,098	1,073	938
Earning from the above loan	--	335	766	1,048
Guarantees received--contra	--	--	696	988
	18,909	22,869	27,750	3,416

	1972	1973*
Fixed assets after depreciation		
Pipeline--natural gas (IGAT)	42,273	--
Pipeline	10,542	13,286
Refineries	13,270	20,513
Marketing	1,837	2,566
Production and exploration	61	23,427
Administration, etc.	571	627
	69,554	60,419
Movable assets after depreciation		
Pipelines	52	64
Refineries	16	151
Marketing	733	628
Production and exploration	140	10,836
Administration, etc.	105	92
	1,046	11,771
Investments and loans		
Investments in subsidiary and		
affiliated companies	23,867	24,666
Long-term loans	3,138	3,463
	27,005	28,129
Current assets		
Stocks of store	2,267	5,939
Stocks of oil	1,410	1,554
Debtors	20,034	40,566
Cash in hand and at bank	11,056	25,714
	34,767	73,773
Long-term loans		
Foreign banks for marketing	3,000	2,709
Second party (IMINOCO J.V. Agreement)	3,274	2,311
Others	1,644	2,281
	7,918	7,301

*(1) Assets amounting to the value of 46,209 million rials have been transferred to the National Iranian Gas Company and Ahwaz Pipe Mill Company in respect of expenditure incurred on the Iran Gas Trunk Line (IGAT) and Ahwaz Pipe Mill Projects and as a result both assets and liabilities figures have been reduced by the above amount.

(2) In accordance with the terms of the 1973 Sales and Purchase Agreements, fixed and movable assets to the value of 35,937 million rials have been incorporated into the accounts of the National Iranian Oil Company in respect of the Abadan Refinery and Fields Areas.

(3) Assets and liabilities in other currencies have been converted to rials at the rates ruling at that date.

(4) To facilitate a further understanding of these accounts it should be noted that 67.50 rials = U.S. $1.00.

Source: National Iranian Oil Company.

TABLE A.5

NIOC's Income Statement, 1966-73
(millions of rials)

	1966	1969	1972	1973
Sales and other income				
Sales of oil products	17,895	25,162	35,066	42,375
Deduct: excise tax, etc.	6,235	8,676	12,219	14,408
	11,660	16,486	22,847	27,967
Sales of crude oil	--	--	--	15,802
Net sales	11,660	16,486	22,847	43,769
Other	257	620	1,496	1,816
	11,917	17,106	24,343	45,585
Cost of oil and operating expenses				
Cost of oil, transportation,				
refining, etc.	6,899	8,723	10,604	15,639
Operating expenses	3,024	2,610	5,203	6,926
Depreciation	1,222	2,780	2,950	3,425
Interest	119	839	431	320
	11,264	14,952	19,188	26,310
Trading profit	653	2,154	5,155	19,275
Add: net income from affiliated				
oil companies	1,348	3,301	8,120	12,750
Net profit	1,991	5,455	13,275	32,025
Add: income from stated payments,				
royalties, bonus, etc.	11,080	17,121	40,029	51,757
Balancing margin	--	--	--	11,142
Net income before taxation	13,071	22,576	53,304	94,924
Appropriations				
Taxation	931	2,583	6,567	16,782
Payment to government against				
stated payments, royalties, and				
balancing margin	10,858	16,778	39,152	61,526
Capital and general reserves	959	2,215	5,290	8,328
Dividend	323	1,000	2,295	8,288
	13,071	22,576	53,304	94,924
Source of funds				
Net income	13,071	22,576	53,304	94,924
Depreciation	1,222	2,780	2,950	3,425
Loans for capital projects, etc.	2,623	4,442	1,887	1,116
Funds received from Plan				
Organization for capital projects	453	14,189	13,228	11,379
	17,369	43,987	71,369	110,844
Application of funds				
Capital projects and investments	3,652	21,414	16,568	18,457
Payment and repayment of loans	614	3,272	3,793	1,043
Taxation	931	2,583	6,567	16,782
Stated payments and balancing				
margin paid to the government	10,858	16,778	39,152	61,526
Dividend	323	1,000	2,295	8,288
Increase in working capital	991	-1,060	2,994	4,748
	17,369	43,987	71,369	110,844

Source: National Iranian Oil Company.

TABLE B.1

Approximate Conversion Factors for Crude Oil*

From \ Into	Metric Tons	Long Tons	Short Tons	Barrels	Kiloliters (cu. meters)	1,000 Gallons (Imp.)	1,000 Gallons (U.S.)
				Multiply by			
Metric tons	1.000	0.984	1.102	7.33	1.160	0.256	0.308
Long tons	1.106	1.000	1.120	7.45	1.180	0.261	0.313
Short tons	0.907	0.893	1.000	6.65	1.050	0.233	0.279
Barrels	0.136	0.134	0.150	1.00	0.159	0.035	0.042
Kiloliters (cu. meters)	0.863	0.849	0.951	6.29	1.000	0.220	0.264
1,000 gallons (imp.)	3.910	3.830	4.290	28.60	4.550	1.000	1.201
1,000 gallons (U.S.)	3.250	3.190	3.580	23.80	3.790	0.833	1.000

To Convert	From			
	Barrels to Metric Tons	Metric Tons to Barrels	Barrels/Day to Tons/Year	Tons/Year to Barrels/Day
	Multiply by			
Crude Oil*	0.136	7.33	49.8	0.0201
Gasoline	0.118	8.45	43.2	0.0232
Kerosene	0.128	7.80	46.8	0.0214
Gas/diesel	0.133	7.50	48.7	0.0205
Fuel oil	0.149	6.70	54.5	0.0184

*Based on world average gravity (excluding natural gas liquids).

Source: British Petroleum Company, Statistical Review of the World Oil Industry (1973), p. 24.

TABLE B.2

Approximate Calorific Equivalents

One million tons of oil equals approximately:	Heat units and other fuels expressed in terms of million tons of oil	Million Tons of Oil*
Heat Units		
41 million million BTUs	10 million million BTUs approximates to	0.24
415 million therms	100 million therms approximates to	0.24
10,500 Teracalories	10,000 Teracalories approximates to	0.95
Solid Fuels		
1.5 million tons of coal	1 million tons of coal approximates to	0.67
4.9 million tons of lignite	1 million tons of lignite approximates to	0.20
3.3 million tons of peat	1 million tons of peat approximates to	0.30
Natural Gas (1 cu. ft. = 1,000 BTUs) (1 cu. meter = 9,000 kcals)		
1.167 thousand million cu. meters	1 thousand million cu. meters approximates to	0.86
41.2 thousand million cu. ft.	10 thousand million cu. ft. approximates to	0.24
113 million cu. ft./day for a year	100 million cu. ft./day for a year approximates to	0.88
Town Gas (1 cu. ft. = 470 BTUs) (1 cu. meter = 4,200 kcals)		
2.5 thousand million cu. meters	1 thousand million cu. meters approximates to	0.40
88.3 thousand million cu. ft.	10 thousand million cu. ft. approximates to	0.11
242 million cu. ft./day for a year	100 million cu. ft./day for a year approximates to	0.41
Electricity (1 kWh = 3,412 BTUs) (1 kWh = 860 kcals)		
12 thousand million kWh	10 thousand million kWh approximates to	0.82

*One million tons of oil produces about 4,000 million units (kWh) of electricity in a modern power station.

Source: British Petroleum Company, Statistical Review of the World Oil Industry (1973), p. 24.

BIBLIOGRAPHY

BIBLIOGRAPHY

Adelman, M. A. "The World Oil Outlook." In National Resources and International Development, edited by M. Clawson. Baltimore: Johns Hopkins Press, 1964.

_____. The World Petroleum Market. Baltimore: Johns Hopkins Press, 1971.

Aghdaii, A. A Forecast of the Iranian Oil Consumption and Its Sources of Supply. Tehran: NIOC Public Relations Office, 1968. In Persian.

Amuzegar, J. Technical Assistance in Theory and Practice: The Case of Iran. New York: Praeger Publishers, 1966.

_____, and A. Fekrat. Iran: Economic Development Under Dualistic Conditions. Chicago: University of Chicago Press, 1971.

Anglo-Persian (Iranian) Oil Company. Annual Report and Balance Sheets, 1909-51.

Baldwin, G. B. Planning and Development in Iran. Baltimore: Johns Hopkins Press, 1967.

Bharier, J. Economic Development in Iran, 1900-1970. London: Oxford University Press, 1971.

_____. "A Note on the Population of Iran 1900-1966." Population Studies 22, no. 2 (July 1968).

British Petroleum Company. Our Industry Petroleum. London, 1970.

_____. B.P. Statistical Review of the World Oil Industry. London, 1971.

Churchill, Winston. The World Crisis, 1911-1914. London: 1923.

The Consortium. Annual Reports of the Iranian Oil Operating Companies, 1955-71.

Dasgupta, B. The Oil Industry in India. London: Frank Cass, 1971.

Economist Intelligence Unit. "Iran," Annual Supplements.

Engler, R. The Politics of Oil. New York: Macmillan, 1961.

Erfani, E. "A Comparative Analysis of Oil Agreements." Tahgigat
Egtesadi: The Quarterly Journal of Economic Research 7
(1970).

Ettehadieh, E. "The Haulage of Petroleum Products in Iran."
Bulletin of Iranian Institute of Petroleum 1 (1957). In Persian.

Farkhan, H. "A Forecast of the Oil Product Demand in Iran." In
A Journey Through the Iranian Oil Industry. Tehran: NIOC
Public Relations Office, 1970. In Persian.

Fatemi, N. S. A Diplomatic History of Persia. New York: Whitter
Books, 1952.

_____. Oil Diplomacy. New York: Whitter Books, 1954.

Fesharaki, F. "The Development of Iranian Oil Industry 1901-71."
Ph.D. dissertation, University of Surrey, March 1974.

_____. "Some Thoughts on the Comparative Evaluation of Oil
Contracts." Tahgigat Egtesadi: Quarterly Journal of Economic Research 9 (1972).

Ford, A. W. The Anglo-Iranian Dispute of 1951-52. Berkeley:
University of California Press, 1954.

Frankel, P. H. Mattei, Oil and Power Politics. London: Faber
and Faber, 1966.

Froozan, M. "Erap-Type Versus Fifty-Fifty Agreements."
Tahgigat Egtesadi: Quarterly Journal of Economic Research 7
(1970).

_____. "Oil Agreements Profitability Comparisons." Iran Oil
Journal, June 1966.

_____. Petroleum Economics. Tehran, 1963. In Persian.

_____, and M. Shirazi. "The Development of Gas Industry in
Iran." Paper presented to the 11th International Gas Congress,
Moscow, June 1970.

Ganji, M. Public International Law. Vol. I. Tehran: 1968. In
    Persian.

Ghassemzadeh, M. An Analytical Comparison of 1933 and 1954
    Agreements. Tehran: 1968. In Persian.

Hartshorn, J. E. Oil Companies and Governments. London: Faber
    and Faber, 1967.

Hirst, D. Oil and Public Opinion in the Middle East. London:
    Faber and Faber, 1966.

Hurewitz, C. Diplomacy in the Near and Middle East--A Documen-
    tary Record, 1914-1956. Princeton: Princeton University
    Press, 1956.

Iran. Annual Household Budget Survey of Iran. Tehran: Bank
    Markazi Iran. Unpublished. In Persian.

_____. Annual Reports and Balance Sheet of the Bank Markazi
    Iran.

_____. Annual Reports of the Iranian Customs and Excise Office,
    1910-51.

_____. Bulletins of the Iranian Institute of Petroleum.

_____. An Industrial Guide to Iran. Tehran: Ministry of
    Economy, 1968.

_____. A Survey of Iranian Economy. Tehran: Iran Trade
    Centre, 1969.

Issawi, C., and M. Yeganeh. The Economics of Middle Eastern
    Oil. New York: Praeger Publishers, 1962.

Kirk, G. "The Middle East, 1945-1950." Survey of International
    Affairs. London: Oxford University Press, 1954.

Leeman, W. The Price of Middle East Oil. Ithaca, N.Y.: Cornell
    University Press, 1962.

Little, I., T. Scitovsky, and M. Scott. Industry and Trade in Some
    Developing Countries--A Comparative Study. London: OECD
    Development Centre, Oxford University Press, 1970.

312   THE IRANIAN OIL INDUSTRY

Longrigg, S. H. Oil in the Middle East. London: Royal Institute of International Affairs, Oxford University Press, 1961.

Looney, R. E. The Economic Development of Iran: A Recent Survey with Projections to 1981. New York: Praeger Publishers, 1973.

Meier, G. M., ed. Leading Issues in Economic Development. New York: Oxford University Press, 1970.

Mikdashi, Z. The Community of Oil Exporting Countries--A Study in Governmental Co-operation. London: Allen and Unwin, 1972.

_____. A Financial Analysis of Middle Eastern Oil Concessions, 1901-1965. New York: Praeger Publishers, 1966.

Mikesell, R., ed. Foreign Investment in Petroleum and Mineral Industries. Baltimore: Johns Hopkins Press, 1971.

Mina, P. "Changes in the Principle of Oil Contracts." Iran Oil Journal, February 1969.

_____. "A Comment on Mr. Stauffer's Paper." Tahgigat Egtesadi: Quarterly Journal of Economic Research 7 (1970).

_____ and F. Najmatabadi. "A Case Study: NIOC's Handling of a Marine Seismic Survey." Iran Newsletter No. 85 (August 1965).

Missen, D. Iran--Oil at the Service of a Nation. London: Transorient Books, 1969.

Modir, M. "The Petroleum Act and the Petroleum Districts." Iran Newsletter No. 73 (August 1964).

Mossadeghi, M. Ahwaz Pipe Mill. Tehran, NIOC Public Relations Pamphlet, 1970. In Persian.

Mostofi, B. "A Review of the History of Oil Development in Iran." Bulletin of the Iranian Institute of Petroleum, November 1964.

Movahed, M. A. Our Oil and Its Legal Problems. Tehran: 1970. In Persian.

Mughraby, M. Permanent Sovereignty Over Oil Resources: A Study of Middle East Oil Concessions and Legal Change. Beirut: Middle East Research and Publishing Centre, 1966.

Naamati, G. A Social and Economic Investigation of the Effects of
     Oil Product Distribution in Iran. Tehran: 1968. In Persian.

Nahai, M., and C. L. Kimbell. The Petroleum Industry of Iran.
     Washington, D.C.: U.S. Department of the Interior, Bureau
     of Mines, 1963.

National Iranian Oil Company. Annual Reports.

_____. Distribution Department Annual Reports.

_____. Getting to Know the Iranian Oil Industry. Tehran: NIOC
     Public Relations Office, 1971. In Persian.

_____. The IGAT Project. Tehran: NIOC Public Relations
     Office, 1971.

_____. The Iranian Oil Industry During the Pahlavi Dynasty.
     Tehran: NIOC Public Relations Office, 1971. In Persian.

_____. A Journey Through the Iranian Oil Industry. Tehran:
     NIOC Public Relations Office, 1970. In Persian.

_____. Management Statistics. Tehran: NIOC Public Relations
     Office, 1972. In Persian.

_____. Oil and Life. Tehran: NIOC Public Relations Office,
     1970. In Persian.

_____. Research in the Iranian Oil Industry. Tehran: NIOC
     Public Relations Office, 1970. In Persian.

_____. Six Decades of the Iranian Oil Industry. Tehran. Undated.

Nemazee, M., and S. Nakazian. "Iran Presents Her Case for
     Nationalization." Oil Forum, 1954.

Nezam-Mafi, M. The Impact of Oil on the Iranian Economy.
     Tehran: NIOC Public Relations Office, 1966. In Persian.

Odell, P. Oil and World Powers. London: Penguin Books, 1970.

Organization of Petroleum Exporting Countries. OPEC Selected
     Documents, OPEC Annual Statistical Bulletin, and various
     other publications.

H.I.M. M. R. Pahlavi. Mission for My Country. New York:
    McGraw-Hill Book Company, 1961.

_____. Shahanshah of Iran on Oil-Background and Perspectives.
    (Interviews and Press Conferences). London: Transorient
    Books, 1971.

Paran, V. "Highlights of Exploration in Iran." Iran Newsletter No.
    96 (July 1966).

Penrose, E. The Growth of Firms, Middle East Oil and Other Essays.
    London: Frank Cass, 1971.

_____. The Large International Firm in Developing Countries--
    The International Petroleum Industry. London: Allen and
    Unwin, 1968.

Plan Organization. The Five Development Plans of Iran, and the
    Appraisal of the Second and Third Plans.

_____. Statistical Yearbook of Iran. Tehran: 1970.

Prebisch, R. "Towards a New Trade Policy for Development."
    Report by the Secretary General of UNCTAD, 1964.

Rouhani, F. A History of OPEC. New York: Praeger Publishers,
    1971.

Saadat, F. An Economic Geography of Iranian Oil. Vols. I and II.
    Tehran: 1966. In Persian.

_____, and A. H. Amini. An Economic Geography of Iran.
    Tehran: 1971. In Persian.

Sadri, M. The Impact of Oil on Iranian Economy. Tehran: NIOC
    Public Relations Office, 1968. In Persian.

Sanghavi, R. Iran, Destiny of Oil. London: Transorient Books, 1971.

Schurr, S. H. The U.S. Oil Conservation. New York: Elsevier Pub-
    lishers, Resources for Future Inc., 1968.

_____, and P. T. Homan. Middle Eastern Oil and the Western
    World--Prospects and Problems. New York: Elsevier Pub-
    lishers, Resources for Future Inc., 1971.

Shwadran, B. The Middle East, Oil and the Great Powers. New
    York: Council for Middle Eastern Affairs, 1959.

Stanford Research Institute. "Long-Term Energy Forecast for
    Iran." Unpublished (1970).

Stauffer, T. R. The ERAP Agreement--A Study in Marginal Taxa-
    tion Pricing. Presented to the Sixth Arab Petroleum Congress
    in Baghdad, December 1966.

_____. "Oil Money and World Money: Conflict or Confluence?"
    Science 184 (April 19, 1974).

_____. "Price Formation in the Eastern Hemisphere." In Con-
    tinuity and Change in the World Oil Industry, edited by Z.
    Mikdashi. Beirut: Middle East Research and Publishing Cen-
    ter, 1969.

Stocking, G. W. Middle East Oil. London: Penguin Books, 1971.

Sweet-Escott, B. A. C. "Financing Problems of Integrated Oil
    Companies." Paper presented to the Fourth Arab Petroleum
    Congress, November 1963.

Tugendhat, C. Oil: The Biggest Business. London: Eyre &
    Spottiswood, 1968.

Turvey, R. Public Enterprise. London: Penguin Books, 1968.

United Nations. Economic Development in the Middle East, 1945-
    1955. New York: United Nations, 1956.

_____. Public Finance Information Papers. Iran, ST/ECA/
    SER.A/14. New York: United Nations, 1951.

_____, International Labour Office. Labour Conditions in the
    Oil Industry in Iran. Geneva: United Nations, 1950.

Votaw, D. The Six Legged Dog: Mattei and ENI--A Study in Power.
    Berkeley: University of California Press, 1964.

Walters, A. A. "The Allocation of Joint Costs with Demand as Prob-
    ability Distribution." American Economic Review 50 (1960).

Weil, R. Jr. "Allocating Joint Costs." American Economic Review
    58 (1968).